Universitext

$16\underset{=}{^{00}}$

William J. J. Rey

Introduction to Robust and Quasi-Robust Statistical Methods

Springer-Verlag
Berlin Heidelberg New York Tokyo 1983

Dr. William J. J. Rey
Philips Research Laboratory
Av. van Becelaere 2 – Box 8
B – 1170 Brussels, Belgium

AMS-MOS (1980) Classification numbers:
Primary: 62F35
Secondary: 62G05, 65U05

ISBN 3-540-12866-2
Springer-Verlag Berlin Heidelberg New York Tokyo
ISBN 0-387-12866-2
Springer-Verlag New York Heidelberg Berlin Tokyo

Library of Congress Cataloging in Publication Data
Rey, William J. J., 1940-
Introduction to robust and quasi-robust statistical methods.
(Universitext)
Bibliography: p.
Includes index.
1. Robust statistics. I. Title.
QA276.R46 1983 519.5 83-20137
ISBN 0-387-12866-2 (U.S.)

© by Springer-Verlag Berlin Heidelberg 1983
Printed in Germany

Printing and bookbinding: Beltz, Offsetdruck, Hemsbach
2141/3140-543210

To the many number-cruncher users who
need trustworthy statistical results

et à Martine,
en espérant qu'elle pourra trouver
plus fréquemment son mari libre.

My own education in Robustness was through the laborious path of extensive reading, applied research and statistical consulting. Particularly during the last decade of fight against errors in data sets, I learned to cope with the observed information, trying to make the best of it. I learned, and have to thank the many colleagues who contributed to my progress; among them I must specially mention those who provided me with challenging problems and those who tolerated my search for more reliable solutions (even if they were not what they expected). Most certainly I also have an important tribute to the worldwide scientific community for its endless questioning of the fundamental issues.

In the preparation of this book I had the privilege to benefit of the technical assistance of the Philips Research Laboratories in Brussels and Eindhoven. In this regard, I would like to state my appreciation to Mr P.T. Logtenberg who carefully corrected a draft version in order to improve my broken English. The type-setting is due to Mrs A.-M. De Ceuster and Mrs E. Moes who kept the computer facility under control (TROFF/UNIX run on VAX); their task was supported by the advices of our software experts. It is a pleasure to acknowledge their excellent collaboration.

er.ror /ˈer-ər/ *n* **1 a** : deviation from a code of behavior.

Merriam-Webster (1973)

The word "normal" in normal distribution must not be thought of in its everyday sense of "usual" or "to be expected". Distributions other than the normal are not abnormal... Sometimes the normal distribution is called Gaussian, especially in engineering and physics. In France it is called Laplacean. These names are probably used because the distribution was invented by de Moivre.

Mosteller, Rourke and Thomas (1970, p. 266)

Table of contents

Chapter 1
Introduction and Summary

The term "robustness" does not lend itself to a clear-cut statistical definition. It seems to have been introduced by G. E. P. Box in 1953 to cover a rather vague concept described in the following way by Kendall and Buckland (1981). Their dictionary states:

> **Robustness.** Many test procedures involving probability levels depend for their exactitude on assumptions concerning the generating mechanism, e. g. that the parent variation is Normal (Gaussian). If the inferences are little affected by departure from those assumptions, e. g. if the significance points of a test vary little if the population departs quite substantially from the Normality the test on the inferences are said to be robust. In a rather more general sense, a statistical procedure is described as robust if it is not very sensitive to departure from the assumptions on which it depends.

This quotation clearly associates robustness with applicability of the various statistical procedures. The two complementary questions met with can be expressed as follows: first, how wide is the field of application of a given statistical procedure or, equivalently, is it robust against some departure from the assumptions? Second, how should we design a statistical procedure to be robust or, in other terms, to remain reliable in spite of possible uncertainty in the available information?

1.1. History and main contributions

With the appearance of involved analytical as well as computational facilities, the field of robustness has received an increased attention in the past thirty years. Mainly, progresses in non-linear mathematics and in iterative algorithms have permitted the new developments. However, robustness stands in the line of many old studies. For instance, a mode can be looked upon as a robust estimate of location, as it also was some twenty four centuries ago. Thucydides relates:

> During the same winter (428 B. C.), the Plataeans... and the Athenians who were besieged with them planned to leave the city and climb over the enemy's walls in the hope that they might be able to force a passage...
> They made ladders equal in height to the enemy's wall, getting the measure by

counting the layers of bricks at a point where the enemy's wall on the side facing Plataea happened not to have been whitewashed. Many counted the layers at the same time, and *while some were sure to make a mistake, the majority were likely to hit the true count*, especially since they counted time and again, and, besides, were at no great distance, and the part of the wall they wished to see was easily visible. The measurement of the ladders, then they got at in this way, reckoning the measure from the thickness of the bricks.

Eisenhart (1971) has also compiled other examples of more or less sophisticated procedures to estimate a location parameter for a set of measurements.

Historical background can also be gained by the account of the state of affairs about 1900 given in Stigler (1973). Briefly, it appears that the adoption of least squares techniques was regarded as second best to unmanageable approaches. The "dogma" of normality was widely accepted; i. e. observations which were in disagreement with the dogma were considered to be erroneous and had to be discarded.

On a donc fait une hypothèse, et cette hypothèse a été appelée *loi des erreurs*. Elle ne s'obtient pas par des déductions rigoureuses; plus d'une démonstration qu'on a voulu en donner est grossière, entre autres celle qui s'appuie sur l'affirmation que la probabilité des écarts est proportionnelle aux écarts. Tout le monde y croit cependant, me disait un jour M. Lippmann, car les expérimentateurs s'imaginent que c'est un théorème de mathématiques, et les mathématiciens que c'est un fait expérimental. - H. Poincare (1912)

(Therefore an hypothesis was made, and that hypothesis has been called *law of errors*. It is not derived by strict deductions; rather crude demonstrations have been advanced and, among the proposals, the one relying on the statement that the probability of a deviation is proportional to the deviation. Nevertheless everybody trusts it, M. Lippmann told me one day, for experimenters believe it is a mathematical theorem, whereas the mathematicians see it as an experimental fact.)

At that time, an important effort was made in the investigation of various rejection procedures as well as in the assessment of their soundness. A remarkable report of this history up to the present time can be found in the series of papers written by Harter (1974-1976). With regard more specifically to robustness history, motivations and theoretical developments are surveyed in the 1972 Wald Lecture of Huber (1972-1973) and complementarily by Hampel (1973a, 1978a). We will deal with them later on.

To gain some more insight in what is robustness we now sketch some important works of the recent past:

- The investigations of von Mises (1947) and Prokhorov (1956) have clarified the relations which exist between asymptotic theory of estimation and finite size practical situations. They provide a neat framework in which the theory of robustness has been able to grow. Particularly, they provide justifications for the analytical derivations.

- A mathematical trick proposed by Quenouille (1956) and augmented by Tukey (1958), the jackknife technique, permits us to reduce the bias and to estimate the variance of most estimators regardless of the distribution underlying the data set. The statistician is thus relieved from some frequently questionable distribution assumptions.

- A review of the principles underlying the rejection of outliers by Anscombe (1960) has stimulated theoretical and experimental research on how to take into account observations appearing in the tails of the sample distributions.

- The famous paper by Huber, in 1964, has been at the origin of answers to the question as to how to design robust statistical procedures. He considers domains of distributions; we excerpt:

> A convenient measure of robustness for asymptotically normal estimators seems to be the supremum of the asymptotic variance ($n \to \infty$) when F (the distribution of the sample) ranges over some suitable set of underlying distributions.

Further, he introduces M estimators and characterizes a most robust family among them for the estimation of the location, when the underlying distribution is a contaminated normal. In this context, we find the estimators to be obtained through minimization of some power p of residuals; the classical least squares estimation (p = 2) have here their places, the pioneering work of Gentlemen (1965) is referred to for other p-values. The maximum likelihood estimators are also M estimators.

- A thesis by Hampel (1968), related to von Mises' work, introduces the influence curve as a tool exhibiting the sensitivity of an estimator to the observation values. Thus, it permits us to modify estimation

3

procedures in order not to depend on outliers, or on any other specific feature of the observations.

- As far as the literature of the period before 1970 is concerned, the interested reader is referred to the annotated bibliography prepared under arrangement with the U. S. National Center for Health Statistics, (N. C. H. S. -1972). Robertucci (1980), Huber (1981) and Hampel et al. (1982) refer to the more recent times.

- Some fresh air has been brought to the field by the two (draft) papers of Efron (1979a, 1979b). They present "bootstrap methods" and their relations with the jackknife technique. Possibly the many presented conjectures will give a new impetus to statistics.

- To conclude the list of main theoretical contributions, it may be worth stressing that many fundamental questions remain open. Among them, the problem of what we really estimate is approached by Jaeckel (1971), and, whether estimators are admissible is analyzed by Berger (1976a, 1976b).

- Complementary to the theoretical progress, experience has been gained through Monte Carlo computer runs. In this respect, it is particularly the Princeton Study (Andrews et al. - 1972) that is worth mentioning because it has clearly shown the need for robust statistical procedures.

1.2. Why robust estimations?

It seems to this author that the main motivation for making use of robust statistical methods lies in the overwhelming power of our computing facilities. In short, it is so easy to perform statistical analyses with computers that, rather frequently, data sets are processed by unsuitable software. Comparison with results produced by robust methods then brings to light any deficiency in the data sets as well as the limitations of the applied statistical procedures. Thus, it appears that robustness is essential because it is complementary to classical statistics. Both of them must contribute to the elaboration and to the validation of statistical conclusions.

4

Many statisticians do not know how poor the methods they apply can be when the data sets do not strictly satisfy the assumptions. This has been brilliantly illustrated by Tukey (1960) in his study on contaminated normal distribution. We recall his findings and supplement them with our own observations. He compares estimations of location and scale performed with a sample drawn from a purely normal distribution $N(\mu, \sigma^2)$ and with a sample drawn from the same normal distribution but contaminated by extraneous observations that are also normally distributed. His model of contaminated normal distribution has the form

$$(1 - \varepsilon)\, N\,(\mu, \sigma^2) + \varepsilon\, N\,(\mu_1, \sigma_1^2).$$

The symmetry is retained for $\mu_1 = \mu$ and, for moderate ε, the contaminated model has shorter tails than those of the normal when $\sigma_1 < \sigma$ and longer ones for $\sigma_1 > \sigma$.

The findings may appear unexpected to some readers. It is well known that estimation of the normal distribution location by the median instead of the mean leads to an efficiency loss of 36 %, but a 10 % contamination is sufficient to have a better estimation by the median. Concerning scale estimation the situation is rather exceptional; a contamination of one or two thousandths is sufficient to balance an efficiency loss of 12 % between the mean deviation and the standard deviation.

In the following two tables, one will find the results obtained for some specific values of parameter ε. Table 1 relates to the symmetric model

$$(1 - \varepsilon)\, N\,(0,1) + \varepsilon\, N\,(0,3^2),$$

whereas the asymmetric model

$$(1 - \varepsilon)\, N\,(0,1) + \varepsilon\, N\,(2,3^2)$$

has resulted in Table 2.

Location has been estimated by the mean and the median. Their respective asymptotic variances (var) constitute a natural basis for

Symmetric model : $(1-\varepsilon)\ N(0,1) +\ \varepsilon\ N(0,9)$

ε	0	0.0018	0.0282	0.1006	0.2436	0.5235	0.8141
Mean							
-	0	0	0	0	0	0	0
var	1	1.0140	1.2252	1.8047	2.9488	5.1882	7.5130
Median							
-	0	0	0	0	0	0	0
var	1.5708	1.5745	1.6315	1.8047	2.2389	3.7066	7.5130
Standard deviation							
-	1	1.0070	1.1069	1.3434	1.7172	2.2778	2.7410
c.v.	0.7071	0.7627	1.1725	1.3540	1.2317	0.9720	0.7929
Mean deviation to the mean							
-	0.7979	0.8007	0.8428	0.9584	1.1866	1.6333	2.0970
c.v.	0.7555	0.7627	0.8514	0.9822	1.0461	0.9720	0.8417
Median deviation to the median							
-	0.6745	0.6754	0.6891	0.7297	0.8259	1.1089	1.6167
c.v.	1.1664	1.1668	1.1725	1.1899	1.2317	1.3435	1.3600
Discriminatory sample size							
0.95	-	7880	497	137	83	123	638

Table 1

comparison.

Scale has been assessed by the three following estimators: the standard deviation, the mean deviation to the mean and the median deviation to the median. A fourth estimator, the semi-interquartile range, has also been computed but has not been reported, the results being the same as (or slightly poorer than) what we obtain with the median deviation to the median. Due to the lack of a natural definition for the scale, it has appeared appropriate to compare the estimators on the basis of their respective coefficients of·variation (c. v.), the ratio

6

of their asymptotic standard deviation to their means.

Asymmetric model : $(1-\varepsilon)\ N(0,1) + \varepsilon\ N(2,9)$

ε		0	0.0008	0.0115	0.0617	0.2228	0.6246	0.6878
	Mean							
$-$		0	0.0016	0.0230	0.1234	0.4456	1.2492	1.3756
var		1	1.0097	1.1377	1.7254	3.4753	6.9348	7.3613
	Median							
$-$		0	0.0005	0.0072	0.0401	0.1657	0.7409	0.8986
var		1.5708	1.5727	1.5977	1.7254	2.2901	6.9348	8.7865
	Standard deviation							
$-$		1	1.0049	1.0666	1.3136	1.8682	2.6334	2.7132
c.v.		0.7071	0.7611	1.1694	1.5176	1.2520	0.8172	0.7837
	Mean deviation to the mean							
$-$		0.7979	0.7996	0.8222	0.9301	1.2841	2.0487	2.1355
c.v.		0.7555	0.7611	0.8263	0.9973	1.0525	0.8077	0.7837
	Median deviation to the median							
$-$		0.6745	0.6749	0.6810	0.7115	0.8368	1.4726	1.6093
c.v.		1.1664	1.1666	1.1694	1.1838	1.2520	1.3731	1.3080
	Discriminatory sample size							
0.95		$-$	8760	769	121	51	119	164

Table 2

The last row of the tables contains a "discriminatory sample size" at level 0.95 which is the minimum size the sample must have to find out whether observations are drawn from a strictly normal distribution or from a given contaminated distribution. We have supposed that a likelihood ratio test was performed to test H_0 against H_1, with

$$H_0 = N\ (\mu_\varepsilon, \sigma_\varepsilon^2)$$

and

7

$$H_1 = (1 - \varepsilon)\, N\,(0,1) + \varepsilon\, N(0 \text{ or } 2, 3^2)$$

for given ε, where μ_ε and σ_ε are the mean and the standard deviation relative to H_1. Numerical integrations have provided the size such that a sample drawn from H_0 be attributed to H_0 by the test with a probability of 0.95. It must be noted that our approach is very simple minded; more realistically we should take into account the estimation of the parameters and, then, the discriminatory sample sizes would be even larger.

Inspection of the tables reveals that the sensitivity to the distribution shape is very dependent upon the nature of the estimators. In particular, a slight long-tail contamination is sufficient to seriously impair the efficiencies of the mean and of the standard deviation. Moreover, quite large sample sizes are required to justify the measurement of the scale by the standard deviation rather than by the mean deviation. Some eight thousand observations should be available to guarantee (at level 0.95) that the standard deviation is the most efficient estimator. This fact must be kept in mind.

In a rather provoking paper some justifications to the use of robustness have been laid down by Hampel (1973a). In what follows, we propose a long excerpt:

> What do those "robust estimators" intend? Should we give up our familiar and simple models, such as our beautiful analysis of variance, our powerful regression, or our high-reaching covariance matrices in multivariate statistics? The answer is no; but it may well be advantageous to modify them slightly. In fact, good practical statisticians have done such modifications all along in an informal way; we now only start to have a theory about them. Some likely advantages of such a formalization are a better intuitive insight into these modifications, improved applied methods (even routine methods, for some aspects), and the chance of having pure mathematicians contribute something to the problem. Possible disadvantages may arise along the usual transformations of a theory when it is understood less and less by more and more people. Dogmatists, who insisted on the use of "optimal" or "admissible" procedures as long as mathematical theories contained no other criteria, may now be going to insist on "optimal robust" or "admissible robust" estimation or testing. Those who habitually try to lie with statistics, rather than seek for truth, may claim even more degrees of freedom for their wicked doings.
> Now what are the reasons for using robust procedures? There are mainly two

observations which combined give an answer. Often in statistics one is using a parametric model implying a very limited set of probability distributions thought possible, such as the common model of normally distributed errors, or that of exponentially distributed observations. Classical (parametric) statistics derives results under the assumption that these models were strictly true. However, apart from some simple discrete models perhaps, such models are never exactly true. We may try to distinguish three main reasons for the deviations: (i) rounding and grouping and other "local inaccuracies"; (ii) the occurrence of "gross errors" such as blunders in measuring, wrong decimal points, errors in copying, inadvertent measurement of a member of a different population, or just "something went wrong"; (iii) the model may have been conceived only as an approximation anyway, e. g. by virtue of the central limit theorem.

1.3. Summary

It appears that most of the robust methods are little founded; this fact must be faced and can be partially attributed to some conceptual difficulty. That these methods are required is due to some uncertainty concerning the statistical model, the quality of the available sample as well as, perhaps, uncertainty concerning the issues involved (e. g., selection of an optimal loss function). These factors prohibit any clear definition of the parameter to be estimated and thus make it impossible to compare an estimator with this parameter. In practice, all the estimators will be analyzed with reference to their asymptotic values obtained either for infinite sample sizes or for an infinite population of finite size samples. Furthermore, it is often possible to compare estimators under a given model at the asymptotic level; therefore, we will encounter robust estimators which are consistent with some parameter. However what is estimated when the model assumption is erroneous may be obscure. This is the conceptual difficulty which lies at the root of all robustness principles.

The book is divided into two parts, the first one being more academic and the second more pragmatic. Part A is largely inspired by our previous writing of "Robust Statistical Methods" (1978). However, in trying to benefit from our experience, the presentation has been fully reorganized and the main issues have been stressed whereas technical details have been relegated to secondary places. We shall often refer to this 1978-book.

Chapter 2 is essentially setting the stage (topologies and limit considerations) where most of the argument takes place. All theoretical derivations are illustrated with reference the same real problem (§2.2) selected for its simplicity, although allocating a probability measure to the sample space (a topological product) is hardly feasible. In section 2.3, we will define the metric spaces we need in order to guarantee the convergence of the following limiting processes; the Prokhorov's metric (2.19) is introduced, which is a rather inconvenient tool for comparing multidimensional probability distributions. The end of the chapter is a discussion of the estimators regarded as functionals of probability distributions. More precisely, consider an estimator $\hat{\vartheta}$ of some parameter set ϑ relative to a probability distribution f (x), i. e.

$$\vartheta = \vartheta(f),$$

while estimating we have at our disposal only an empirical distribution g (x) and we may write

$$\hat{\vartheta} = \vartheta(g).$$

This empirical density of probability relates to the sample set $\{\, x_1,...,x_n \,\}$ of size n where each multidimensional observation x_i can be allocated a given weight w_i; this yields

$$g\ (x) = (\textstyle\sum w_i)^{-1} \sum w_i \delta(x - x_i) \tag{2.40}$$

where $\delta(x - x_i)$ stands for the Dirac function concentrated on x_i. Section 2.3 presents the fundamental relation

$$\hat{\vartheta} = \vartheta + \frac{1}{1!(\sum w_i)} \sum w_i \Psi\ (x_i)$$

$$+ \frac{1}{2!(\sum w_i)^2} \sum w_i\ w_j\ \varphi\ (x_i,x_j) \tag{2.41}$$

$$+ \cdots$$

where the "coefficients" $\Psi\ (x_i), \varphi\ (x_i,x_j), \cdots$ are defined with respect to the distribution f (x) and, therefore, are independent of the empirical density. The conditions underlying the expansion (2.41) are very mild and the convergence occurs with the von Mises' differentiable statistical functions. Considering the estimator $\hat{\vartheta}$ as a function of the set of weights, we evaluate the coefficients of (2.41) without referring to the probability density f anymore; for instance we obtain

$$\Psi(x_i) = (\sum w_i)(\partial / \partial w_i) \hat{\vartheta}. \tag{2.48}$$

The concept of robustness is discused in Chapter 3. The breakdown point (§3.2) and the influence function (§3.3) are presented from Hampel's standpoint and it becomes clear that the "coefficient" $\Psi(x_i)$ in (2.48) plays a dominant role; it has been termed "influence curve" or "influence function". Robustness implies a bounding of the influence function.

The jackknife technique is the subject of Chapter 4. Although one might think that this is a minor statistical trick however, it provides very general results when properly applied to the expansion (2.41). It may be used as suggested by Quenouille to reduce estimator bias (4.16), but its contribution to estimating the second and third order moments appears much more interesting. Let us define a variate δ_i by

$$\delta_i = \hat{\vartheta}_i - \hat{\vartheta} \tag{4.18}$$

$$= (n-1) \left[\left(\frac{1}{n} \sum \hat{\vartheta}_i \right) - \hat{\vartheta}_i \right]$$

where $\hat{\vartheta}_i$ is the estimator corresponding with the sample of size (n-1), being the original sample of size n where the observation x_i has been deleted. This variate δ_i is a finite version of another variate obtained by perturbing the weight of x_i rather than deleting this observation; this infinitesimal version of δ_i is equal to the influence function

$$\Psi(x_i) = \delta_i. \tag{4.35}$$

and approximates the finite version (4.18). Either version can be introduced into the estimator of the variance-covariance matrix

$$Cov\ (\hat{\vartheta}) = \frac{1}{n(n-1)}\sum \delta_i\ \delta'_i \qquad (4.31)$$

as well as in the estimator of the third central moment

$$\mu_3\ (\hat{\vartheta}) = \frac{1}{n(n-1)(n-2)}\sum \delta_i^3. \qquad (4.33)$$

The jackknife method can be seen as a simplified bootstrap technique. Such a technique, presented in Chapter 5, has the merit of proposing a way to "observe" the sampling distribution of an estimator even though only a single sample is available. Efron has designed the method and it is based on a random sampling scheme with resubstitution. The principle underlying the bootstrap is worthy of further developments even if the present results do not fulfill the expectations.

The second part of this book deals with the ordinary routine work in statistics; the data sets are often distributed according to some bell shape, however it can seldom be guaranteed that a bell-shape distribution behaves like a normal distribution. Whether little adequacy of the assumptions does matter and how to proceed in order to be little sensitive to the data set ill-conditioning are the main topics of Part B. This has led us to disregard certain asymptotic aspects which otherwise would be of great concern.

Chapter 6 presents the type M estimators. They are generalizations of the ordinary likelihood functions and are such that they minimize a criterion

$$\sum \rho(x_i,\ \vartheta) = \min \text{ for } \vartheta \qquad (6.1)$$

where ρ (.) is a (convex) function in the parameter set ϑ, depending upon the observation x_i. Some simple analytical work leads to the influence function (6.13) as well as to the variance-covariance matrix (6.14) of $\hat{\vartheta}$. In order to guarantee the unicity of the estimator, it is convenient to impose that

$$\rho\ (x,\ \vartheta)\ be\ convex\ in\ \vartheta. \qquad (6.17)$$

A much more difficult problem is imposing robustness; an algorithm is proposed to bound the n-dimensional influence function (6.22) but this approach appears to be too involved for any routine use. This leads us to define **quasi-robust**estimators (§6.4) which are more robust than ordinary estimators without being robust in the strict sense of the theory. Several possibilities can be considered, but an outstanding ρ-function to be introduced into (6.1) is

$$\rho\,(u) = 2c^2\,[\ |u|/c - 1n(1+|u|/c)] \tag{6.27}$$

with

$$c = 0.17602.$$

This function, called Fair, can replace the square function u^2 in all least squares procedures; then the variable u is a residual scaled in order to have a unit standard deviation asymptotically.

Chapter 7 covers L estimators, a class of estimators which are linear functions of the sample quantiles (7.5). To this class belongs the trimmed mean estimator of location (7.13) and, except for details, also the median absolute deviation

$$MAD = median\,[\,|x_i - median\,(x_i)|\,], \tag{7.15}$$

which is an estimator of the scale. We conclude the chapter with observing that MAD can be much improved without impairing its robustness.

Chapter 8 recalls the existence of R estimators, which are the third class of estimators as presented by Huber. They consist in sets of parameters such that a rank test statistics cancels. In spite of their distribution-free merits, it seems that their complexity discourages most people from applying them.

Chapter 9 can be considered as the most important one for applications; it is relative to MM estimators or, in other terms, to M estimators which are defined by a set of simultaneous equations (9.2) rather than by a single equation (6.1). This difference is essential, for it drastically modifies the properties, e. g. they can be quasi-robust while the corresponding M estimators are strictly robust. Investigation of the linear model

$$y_i = z'_i \vartheta + \varepsilon_i \tag{9.9}$$

is thoroughly conducted in order to support the theoretical results of Part A. One of the outcomes is a definition of a scale estimator, s by (9.20), which appears to be the only one consistent with the minimization criterion (6.1). Eventually we obtain that the g-dimensional estimator $\hat{\vartheta}$ is the solution of the two-equation system

$$s^2 \sum \rho \left[(y_i - z'_i \vartheta) / s \right] = \min \text{ for } \vartheta,$$

$$1 = \frac{1}{(n-g)\, Int} \sum \rho \left[(y_i - z'_i \vartheta) / s \right], \tag{9.24}$$

where

$$Int = .187442$$

whenever ρ (.) is the above function Fair. The multidimensional location-scatter problem (9.6) and the non-linear regression models (9.7) are resolved under the same requirement of compatibility between the minimized criterion and the scatter definition. To conclude with, attention is devoted to the available numerical methods, especially to the fixed point algorithms (9.8.3).

Chapter 10 tackles two similarly oriented problems, i. e. setting confidence limits either onto the sample space or into the parameter space. The first problem is resolved in the regression context with the help of quantile estimators. The second problem seems to correspond with an ill-posed problem and, thus, would not have any answer of general validity.

The last chapter covers two subjects which have appeared throughout the previous text without being dealt with. We discuss outliers and their treatment or, rather, why we do not treat them and prefer to handle them without following special procedures. Then we consider analysis of variance and, more generally, the solution of problems with constraints. This topic well indicates that many of our ways

14

of thinking must be revised.

Then a limited bibliography is given. We have tried to favor recent papers and surveys in the domains which are secondary to our main investigation. We are well aware of the fact that such a selection is very arbitrary. Furthermore, we feel limited in readability, time and space.

Throughout this text, we have not devoted much attention to the asymptotic properties and have placed emphasis on the behavior of finite sample estimators. This option results from the need for robust methods in applications.

PART A
The Theoretical Background

Chapter 2
Sample spaces, distributions, estimators ...

2.1. Introduction

In the following sections, we present a large body of theory which still is in its infancy. We could have strived at rigor, generality and completeness but, then, we would have stuck halfway in conjectures. The main issues are not yet clear. For simplicity, we have decided to restrict ourselves to what we need, leaving to the experts the task of extending and improving the main arguments. We hope that, sooner or later, some motivated analysts highly qualified in topology will sustain the investigations by von Mises and Prokhorov. We ourselves resort to the elementary tools of mathematical analysis.

In order to illustrate the theoretical aspects we encounter, we shall discuss the analytical derivations with reference to an example. We deal with the problem of mean-life estimation based on a right-censored sample.

2.2. Example

Suppose we have a set of items which run from an instant of starting to an instant of either stopping or failure. We are interested in estimating the mean time interval between these instants assuming that samples of these items are independently drawn from some common population. These items can be, say, machines, lamp bulbs or people; in the last-mentioned case the instant of stopping would be the one from which these people are no longer observed, whereas failure must be understood to refer to disease or death. For this reason we use the term of mean life.

The main notations are as follows. Given a sample of size N drawn from a population of items with a cumulative distribution function (cdf) of the instant of failure $G(t)$ and probability density function (pdf) g (t), two different types of items are present: type A for the N_A items stopped before failure and type B for the N_B failing items. We shall not enter here into the incidence of the stopping time distribution; this may be found in Rey (1975b). The problem can now be stated more precisely: given a sample of epochs

$$X_N = (t_{1,A}, t_{2,A}, \ldots, t_{N_A,A}, t_{1,B}, \cdots, t_{N_B,B})$$

with

$$N = N_A + N_B ,\tag{2.1}$$

we require an estimator of the mean life

$$\tau = \int_0^\infty t \, d\, G .\tag{2.2}$$

Such a data set is reported in Table 3. It has a sample size $N=15$, with $N_A=6$ items stopped before failure and $M_B=9$ failing items. The reported observations are the number of man-days devoted to various projects in statistical consultation. The projects were started later than July 1, 79 and were observed until June 30, 80.

Life-time Data Set

Time	Type	Reversed Rank	Hazard	Cumulative Hazard
1	A	15.0		
3	B	14.0	0.0714	0.0714
4	B	13.0	0.0769	0.1484
5	A	12.0		
8	B	10.5	0.0952	0.2436
8	B	10.5	0.0952	0.3388
9	B	9.0	0.1111	0.4499
10	B	8.0	0.1250	0.5749
12	A	7.0		
20	A	6.0		
25	B	5.0	0.2000	0.7749
40	B	3.5	0.2857	1.0607
40	A	3.5		
75	A	2.0		
100	B	1.0	1.0000	2.0607

A : Right-censored data, $N_A = 6$, $S_A = 153$.

B : Full life data, $N_B = 9$, $S_B = 207$.

Total, $N = 15$, $S_T = S_A + S_B = 360$.

Table 3

Before proceeding it could be wise to assess by standard graphical methods how the data set looks like. According to Nelson (1972), we rank the observations in reversed order, and subsequently we estimate the "hazard" (or "mortality force") as

$$hazard = 1/(Reversed\ Rank)$$

for type B items and, eventually, we add up these estimates to obtain a non-parametric estimate of the cumulative hazard function

$$H(t) = -\ln[1-G(t)] .$$

These steps are listed in Table 3 and the estimate $H(t)$ is plotted in Fig, 1 on log-log graph paper. It is seen that the data points lie in the vicinity of a straight line with a slope of 45°. This is consistent with an assumption of constant hazard rate, i. e.

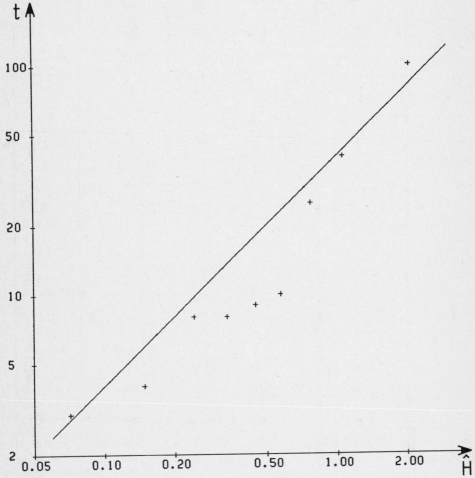

Fig.1. Life-time Data Set. Time t against cumulative hazard function H.

$$G(t) = 1 - e^{-\lambda t} \qquad (2.3)$$

On the basis of the model (2.3) and in virtue of (2.2), the mean life is

$$\tau = \frac{1}{\lambda} . \qquad (2.4)$$

This definition of τ implies that the model (2.3) is assumed to be strictly valid, however in practice we have no method to test such an assumption. This is discussed at length in Rey (1975b). Therefore we slightly modify the statement of the problem and say that we intend to estimate the unknown parameter ϑ, in the model

$$G(t) = 1 - e^{-t/\vartheta} .\tag{2.5}$$

In order to design an estimator of ϑ, we now proceed along the standard techniques. We will meet with Bayesian views and eventually select a maximum likelihood estimator.

The appeal of Bayes's estimation procedure to intuition is such that it largely compensates for many of its drawbacks. Essentially we develop the argument without devoting too much attention on what prior assumptions we should use and, then, we inspect the result and revise our assumptions to be in agreement with common sense. This is in harmony with Ferguson (1974) and Robinson's (1979) views.

We denote probabilities $Pr\{.\}$, by conditional probabilities by $Pr\{.|.\}$ and densities of probability by $p\{.\}$. Then, according to the Bayesian principles of inference (see Malinvaud (1970) Chap. 2, § § 11-12, for a very elegant and concise exposition)

$$p\{\vartheta|X_N\} = \frac{Pr\{X_N|\vartheta\}\ p\{\vartheta\}}{\int Pr\{X_N|\vartheta\}\ p\{\vartheta\}\ d\vartheta} .\tag{2.6}$$

The straightforward part of the derivation is the factor relating a probability with X_N by (2.1) for a given value of the parameter ϑ in the model (2.5). We have

$$Pr\{X_N|\vartheta\} \propto \prod_{i=1}^{N_A} [1 - G(t_{i,A})] \prod_{j=1}^{N_B} g(t_{j,B})$$

$$= \vartheta^{-N_B}\ e^{-S_T/\vartheta} \tag{2.7}$$

where the symbol $\propto$ denotes the proportionality, and

$$S_A = \sum_{i=1}^{N_A} t_{i,A} ,$$

$$S_B = \sum_{j=1}^{N_B} t_{j,B} \, , \qquad (2.8)$$

$$S_T = S_A + S_B \, .$$

As pointed out before, the way in which the prior distribution $p\{\vartheta\}$ should be selected is not obvious. It seems to us natural to assume that $\ln\vartheta$ is uniformly distributed over a very large interval. This leads to the assumption

$$p\{\vartheta\} = \begin{cases} [2\,\vartheta\,\ln c_2]^{-1} \,\, , \text{if } c_1/c_2 < \vartheta < c_1 c_2 \\ 0 \qquad\qquad\quad , otherwise \, . \end{cases} \qquad (2.9)$$

Hence, by (2.6) we obtain the posterior distribution

$$p\{\vartheta\,|\,X_N\} = S_T^{N_B}\,[(N_B-1)!]^{-1}\,\vartheta^{-(N_B+1)}\,e^{-S_T/\vartheta} \qquad (2.10)$$

This formula holds for large c_2 and then does not depend upon c_1 and c_2, the arbitrary parameters in (2.9). The distribution (2.10) can be considered as fiducial for making inferences. The first two moments are

$$E\{\vartheta\} = S_T/(N_B-1) \qquad (2.11)$$

and

$$E\{\vartheta^2\} = S_T^2/\,[(N_B-1)(N_B-2)] \, ,$$

yielding a variance estimate

$$E\{[\vartheta-E\{\vartheta\}]^2\} = S_T^2/[(N_B-1)^2\,(N_B-2)] \qquad (2.12)$$

We have developed these expressions without any clear definition of what we mean by the expectations $E\{.\}$.

The mean value being the best estimator under a quadratic loss function, we may use it as a definition; e. g.

$$\hat{\vartheta} = S_T/(N_B-1) \, . \qquad (2.13)$$

But other formulations could equally well be considered.

The maximum likelihood estimator is found by maximizing (2.7); hence it is

$$\hat{\vartheta} = S_T / N_B \qquad (2.14)$$

Pitman's estimator (1939) also exhibits nice properties of unbiasedness and optimality. In our notation, it takes the form

$$[\int \vartheta \, Pr\{X_N | \vartheta\}] / [\int Pr\{X_N | \vartheta\} \, d\vartheta]$$

or

$$\hat{\vartheta} = S_T / (N_B - 2) \qquad (2.15)$$

Thus far, we have found three competitors and have good arguments to qualify any of them. Taking into account the scatters of these three point estimates, it may be observed from Table 4 that they little differ in spite of the small sample sizes. Table 4 contains the main findings relative to our illustrative data set. The investigation has been specially described for estimator (2.13), only numerical data are produced for (2.14) and (2.15). The Prokhorov distance has been evaluated between the type B observations and the theoretical distribution (2.5); this is reported at the end of § 2.3 (see (2.24) et sq.). A Taylor-like expansion (2.52) will provide further major results: the influence function equations (end of § 3.3), a bias correction by the Quenouille jack-knife as well as moment estimations (see § 4.3) - It will be noted that the standard deviation estimates 18.44 and 18.21 are close to 17.01, the value found from (2.12) - Finally, "failure" of the bootstrap method has been met with (end of §5.1).

2.3. Metrics for probability distributions

As indicated by Munster (1974), we need not to restrict our distributions to Baire functions, nor the application space to Borel sets; although, in practice, we only feel at ease with the Baire class of functions and the Euclidean space R^n. We will be concerned here with methods of measuring the closeness of distributions; the assessment of

22

	Estimator definitions		
	(2.14)	(2.13)	(2.15)
Estimator value $\hat{\theta}$	40.00	45.00	51.43
Bias corrected See (4.39) $\tilde{\theta}$	38.02	36.45	32.03
Standard deviation : σ See (4.40), (4.41) See (4.35), (4.41)	15.53 15.51	18.44 18.21	22.63 22.01
Skewness : $\gamma_1 = \mu_3/\sigma^3$ See (4.40), (4.42) See (4.35), (4.42)	0.208 0.222	0.181 0.199	0.151 0.176

Table 4

closeness between estimators will later be reduced to the closeness of their sampling distributions and, hence, only distributions are of interest here.

The distributions we are concerned with are probability densities defined with respect to the Lebesgue measure. This is not as restrictive as it might appear and alternative choices of the measure are possible in theory. We meet with continuous and discrete (or not everywhere continuous) distributions. The latter distributions will be seen as sums of Dirac functions rather than as smooth functions only defined on some discrete sample space.

Let Ω be the whole metric space and $\omega \subset \Omega$ be some subset, we associate the probability measure $F(\omega)$ to the probability density function f by

$$F(\omega) = \int_\omega f(x) \, dx \; .$$

(2.16)

23

Whenever $f(x)$ is not continuous for any $x \in \omega$, the integral notation must be understood in the sense of the distribution theory.

The sample space Ω can possibly be different for each distribution f, however we do not consider this possibility here and assume, first, that it can be extended in order for it to be common to all distributions f and, second, that it has a unit probability measure,

$$F(\Omega) = 1 . \tag{2.17}$$

For any three functions f, g and h belonging to the distribution space E,

$$f, g, h \in E ,$$

we define a metric by a distance function $d(.,.)$ having the ordinary properties

$$d(f, g) \geq 0$$

$$d(f, g) = 0 <==> f = g$$

$$d(f, g) = d(f, g)$$

$$d(f, g) \leq d(f, h) + d(h, g) .$$

The space is complete or, in other terms, the limit of any Cauchy sequence $\{f_n\}$ of E belongs to E, i. e.

$$f_n \in E ,$$

$$\lim \{f_n\} = f ==> f \in E .$$

The space is convex, i. e.

$$f, g \in E, t \in R, 0 \leq t \leq 1$$

$$h = tf + (1-t)g ==> h \in E .$$

It must be observed that we have as much as possible avoided the axiomatic theory of probability introduced by Kolmogorov, as related by Feller (1966, Chap. 4); this is because we have found it too constrained and needlessly unmanageable. A critical appraisal of the

Kolmogorov set-up has been reported by Fine (1973, Chap. 3) and justifies our view.

A great number of "distance" definitions have been proposed to assess the closeness of distributions. However they are generally not acceptable in our context. In perusing Kanal's review (1974) or Chen's (1976), the following deficiencies are observed for most distances :

- They provide a measure of the closeness of continuous distributions, but, are not appropriate to compare an empirical (discrete) distribution with its underlying parent.

- They may be strictly one-dimensional and, thus, are not applicable to R^n.

- They rarely satisfy the triangular inequality, a condition which is very useful to compare an empirical distribution with a continuous distribution differing from its parent.

To the best of our knowledge the only distance measure suitable to our purpose has been proposed by Prokhorov (§ 1.4, 1965); this is although Dowson and Landau (1982) recommend an extension of the Frechet's distance measure, note that Garcia Polamores and Giné have demonstrated a certain optimality of Prokhorov's standpoint. It is simple in its main idea although rather involved in its details. To ease the understanding of its analytical aspects, we now consider the three points just mentioned. The Prokhorov metric permits comparison of a discrete empirical distribution with a continuous one through the association of each observation of the former with a subset of the sample space; the comparison is then performed with the help of the probability of the latter distribution over this subset. The distance measure is derived from the probability measures achieved over subsets of the sample space and is, hence, independent of dimensionality considerations, In fact, its definition refers to the supremum of probability measures and, accordingly, the triangular inequality holds true.

- Although we have required the possibility of considering simultaneously both types of distributions, we must say that several papers avoid this constraint and hence also avoid the Prokhorov metric. For

25

instance, Beran (1977a, 1977b) first substitutes a continuous distribution to any discrete data set and, then, performs estimation by minimizing the Hellinger distance.

Proposed by Hampel (1971) for assessing robustness, the Prokhorov metric is increasingly used in probability theory to investigate limit theorems - e. g., see Komlos et al. (1975) - It induces a weak-star topology over the space of the distributions, i. e. a topology of pointwise convergence. We first define the Prokhorov metric between two functions f and g over a common sample space Ω. Different sample spaces could be attributed to different distributions, but we assume they can be extended in order to be common to all distributions while satisfying (2.17) for each individual sample space.

In a metric space Ω, to any subset $\omega \subset \Omega$ we associate an open ε-neighborhood ω^ε of all points at a distance less than $\varepsilon\sigma$ from ω. Formally, with $\rho(.,.)$ the metric of the Ω-space,

$$\omega^\varepsilon = \{ y : \exists\, x \in \omega, \rho(x,y) < \varepsilon\sigma \} \tag{2.18}$$

where σ is a given positive constant which will soon be discussed. Then the metric is given as the distance $d(f,g)$ between the distributions f and g, associated with the measures F and G, in the following way

$$d(f,g) = \inf_\omega \{ \varepsilon \geq 0 : F(\omega) \leq G(\omega^\varepsilon) + \varepsilon \}$$

$$= \inf_\omega \{ \varepsilon \geq 0 : F(\omega) \leq F(\omega^\varepsilon) + \varepsilon \} \ . \tag{2.19}$$

The equality of definitions results from (2.17), namely

$$F(\Omega) = G(\Omega) \ .$$

The infimum runs on all subsets ω. A more explicit definition can be stated as follows:

- With any subset ω, we associate $\rho(\omega)$ through the definition

$$\rho(\omega) = \min_\varepsilon \{ \varepsilon : F(\omega) \leq G(\omega^\varepsilon) + \varepsilon \} \ . \tag{2.20}$$

- Then the Prokhorov distance is given by

$$d(f,g) = \max_{\omega} \{\rho(\omega)\} . \qquad (2.21)$$

The converse definition can be obtained by permuting F and G in (2.20).

In order to further explain the concepts, we first consider a situation where the probability densities f and g are everywhere bounded. In such cases, strict equalities are realized in (2.19) and (2.20). An illustration of the components appearing in (2.20) is given in Fig. 2 for $\omega \subset E^1$. Furthermore, it is easy to see that (2.21) implies equal densities on the frontiers of the various sets. We have, while ω_0 is the set yielding the infimum

$$f\,[Fr(\omega_0)] = g\,[Fr(\omega_0^\varepsilon)] \qquad (2.22)$$

This relation is often very useful in deriving the optimal set ω_0. This is shown in Fig. 3 for the two exponential distributions according to (2.3); the two forms of the Prokhorov distance definition (2.19) lead to dual definitions of the optimum set ω_0.

So far we have not paid much attention to the parameter σ appearing in the definition of the ε-neighborhood by (2.18). A first remark, which might be appropriate because Prokhorov (1956) did not introduce such a scale factor, is that a definition with $\sigma = 1$ (a pure number) presents some lack of consistency. Effectively, consider the dimension of ε: (2.18) implies that ε is a distance measure expressed, say, in meters, while (2.19) and (2.20) introduce ε as a probability measure, the difference between two probabilities (not in terms of meters). To remove this inconsistency, we introduce an axiom of linear invariance as also did Johnson (1979) : the Prokhorov distance should remain insensitive to linear transformations of the sample space Ω. To obtain this insensitivity, we introduce the scale factor σ; it has the dimension of a distance in the metric space Ω.

The Prokhorov distance is usually difficult to estimate. This will be clear in the two following situations, both relating to the exponential distribution (2.3). The difficulty can be attributed to the fact that the

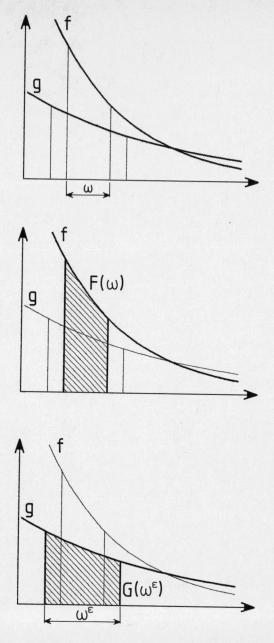

Fig.2. Prokhorov's Metrics. Definition of the components appearing in (2.20).

resulting analytic equations happen to be non-linear in most cases.

Let us derive the Prokhorov distance between the two one-dimension distributions

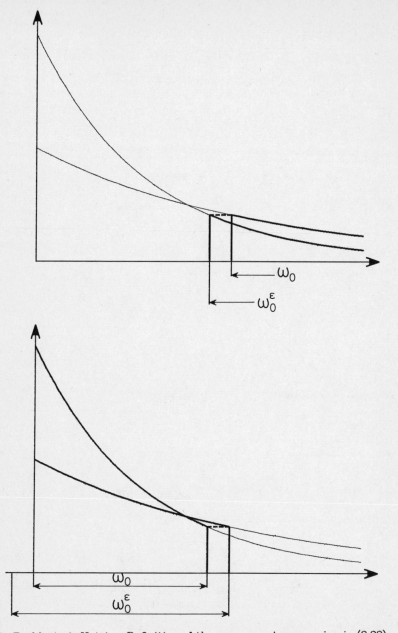

Fig.3. Prokhorov's Metrics. Definition of the components appearing in (2.22).

$$F(x) = 1 - e^{-\lambda_1 x}$$

and

$$G(x) = 1 - e^{-\lambda_2 x} \qquad (2.23)$$

with

$$\lambda_1 \leq \lambda_2$$

The situation is represented by the upper graph of Fig. 3 for which

$$\omega_0 = [x,\infty]$$

and

$$\omega_0^\varepsilon = [x-\sigma\varepsilon,\infty] .$$

(2.19) and (2.22) provide the fundamental equations

$$F(\omega_0) = G(\omega_0^\varepsilon) + \varepsilon$$

and

$$f(x) = g(x-\sigma\varepsilon) ,$$

with

$$d(f,g) = \varepsilon .$$

Hence, we must solve for ε the non-linear system of equations

$$\exp(-\lambda_1 x) = \exp[-\lambda_2(x-\sigma\varepsilon)] + \varepsilon ,$$

$$\lambda_1 \exp(-\lambda_1 x) = \lambda_2 \exp[-\lambda_2(x-\sigma\varepsilon)] .$$

Algebraic elimination of x between the two equations yields the implicit solution

$$\lambda_1\lambda_2 \, \sigma \, \varepsilon = \lambda_1 \, \ln(\lambda_1/\lambda_2) + (\lambda_2-\lambda_1) \ln\left[(\lambda_2-\lambda_1) / (\lambda_2 \, \varepsilon)\right] .$$

In order to clearly see the dependency of this equation on σ, we introduce the two intermediate variables u and v, such that u be independent of σ as well as satisfying

$$\varepsilon = u \, e^{-v\sigma} .$$

It comes

$$\lambda_1 \ln(\lambda_1/\lambda_2) + (\lambda_2-\lambda_1) \ln[(\lambda_2-\lambda_1)/(\lambda_2 u)] = 0$$

and

$$\lambda_1\lambda_2 u\, e^{-v\sigma} = (\lambda_2-\lambda_1) v \ .$$

The first equation has the exact solution

$$u = \frac{\lambda_2-\lambda_1}{\lambda_2}\, \exp\left(\frac{\lambda_1}{\lambda_2-\lambda_1}\, \ln\frac{\lambda_1}{\lambda_2}\right)$$

or, approximately,

$$u \approx (\lambda_1-\lambda_2)\, e^{-1}\, [\,\frac{1}{2}(\frac{1}{\lambda_1} + \frac{1}{\lambda_2})]\ .$$

The second equation is transformed into a quadratic one with the help of the very good approximation

$$\ln y \approx \frac{1}{2}(y - \frac{1}{y})$$

Hence the result

$$v = a(2\sigma a+1)^{-1/2}$$

where

$$a = \lambda_1 \exp\left(\frac{\lambda_1}{\lambda_2-\lambda_1}\, \ln\frac{\lambda_1}{\lambda_2}\right)$$

$$\approx e^{-1}\, [(\lambda_1+\lambda_2)/2]$$

Finally, from these items, we obtain the distance of Prokhorov which is valid under conditions

$$\lambda_1 \approx \lambda_2\ ,$$

$$\lambda^*\sigma = 1\ ,$$

$$\frac{1}{\lambda^*} = \frac{1}{2}(\frac{1}{\lambda_1} + \frac{1}{\lambda_2})\ .$$

Its approximate value is

$$d(f,g) \approx .27825 \ |\lambda_2 - \lambda_1| / \lambda^* ,$$

and a much more correct approximate value has been obtained by numerical means, it is

$$d(f,g) \approx .27846 \ |\lambda_2 - \lambda_1| / \lambda^* .$$

A comparison between the last two expressions demonstrates the quality of the approximations.

So far we have been mainly concerned with distances between continuous distributions. A second situation of interest is what happens when one of the two distributions is discrete (in what follows the distances between two discrete distributions will not be examined). The easiest way to evaluate the Prokhorov distance is to consider that the discrete distribution is the limit of a sequence of smooth distributions - see Abramov (1976) and Richardson (1980). We may for instance associate a smooth kernel with each discontinuity. We illustrate the procedure on the § 2.2 example.

We consider the empirical distribution representative of the type B observations in Table 3, namely

$$f(t) = \frac{1}{m} \sum \delta (t - t_i) \tag{2.24}$$

where

$$m = N_B = 9$$

$$t_i = t_{iB}$$

and $\delta(.)$ stands for the Dirac distribution. The distribution of comparison will be (2.5) or, equivalently,

$$g(t) = \frac{1}{\vartheta} e^{-t/\vartheta} \tag{2.25}$$

Then, to estimate the distance, we must decide that σ will be introduced into (2.18). We have selected the very natural

$$\sigma = \vartheta = observed \ standard \ deviation . \tag{2.26}$$

This selection not only sets σ, but also further defines the distribution of comparison (2.25).

We associate a member of the distribution family $\{h_n(.)\}$ with each Dirac kernel such that

$$\lim_{n \to \infty} h_n(.) = \delta(.) \tag{2.27}$$

with point-wise convergence. For instance, the distributions could be members of the normal family

$$h_n(x) = (n/\sqrt{2\pi}) \exp(-n^2 x^2/2) .$$

Hence, we are interested in $d(f_n, g)$ related to the desired Prokhorov distance by

$$d(f, g) = \lim_{n \to \infty} d(f_n, g)$$

$$f_n(t) = \frac{1}{n} \sum h_n (t - t_i) .$$

We now define the set ω_0 yielding the infimum in the first definition (2.19). Inspection of the inequalities reveals that ω_0 is the union of some sets ω_i (possibly all sets ω_i) defined according to

$$\omega_i = \{t : h_n (t - t_i) > g(t)\} .$$

Indeed, any part of ω_0 not belonging to the ω_i, namely

$$\omega_0 \cap [(\cup \omega_i) ,$$

contributes to the right hand side of (2.20) without increasing its left-hand side. Hence if not empty, it yields a reduced $\eta(\omega)$. This demonstrates that left-hand

$$\omega_0 \cap [(\cup \omega_i) = \phi .$$

Now consider the set of the ω_i -union

$$\omega = \cup \omega_i .$$

which corresponds to some ε. Delete one of the ω_i; it is clear that the decrease of $F(\omega)$ might be smaller than the decrease of the $G(\omega^\varepsilon)$ -term. This yields a larger $\eta(\omega)$ and, thus, a better set ω. In view of these considerations, we have proved that

$$\omega_0 = \bigcup (selected\,\omega_i).$$

Inasmuch as our illustrative case is concerned, one obtains

$$\sigma = \vartheta \approx 31.22$$

$$\omega_0 = \{\, 3, 4, 8, 9, 10, 100 \,\}$$

$$d(f,g) \approx .294342.$$

It will be noticed that the two observations 25 and 40 do not appear in the ω_0 set. As a matter of fact the union of all ω_i is not the optimal set; it yields

$$\eta(\bigcup \omega_i) \approx .227044.$$

2.4. Estimators seen as functionals of distributions

In a paper on the asymptotic properties of sample distributions, von Mises (1947) introduces in a heuristic way some Taylor-like expansion of estimators in terms of the distribution of the underlying sample. This expansion is valid in a non-clearly defined context of "differential statistical functions". In what follows we shall present this theory and provide an original delimitation of the field of application. But let us first introduce von Mises's argument.

The basic material consists of two distributions f and g, which can be viewed as points in a convenient distribution space, and an estimator, or rather a functional, on these distributions; let it be $T(f)$ or $T(g)$. The estimators are defined over some sample space we assume to be common to both and which could be discrete as well as continuous, thus far. We are now interested in the difference between the two functionals $T(f)$ and $T(g)$.

Consider the functional $T(h)$ where h is a distribution intermediate between f and g, thus

$$h(x) = (1-t)\, f(x) + t\, g(x)$$

with

$$0 \le t \le 1.$$

Under suitable conditions, this functional is a continuous function of the real variable t and can be expanded in Taylor series:

$$T(h) = T(f) + t\, a_1 + (t^2/2)a_2 + \cdots,$$

which converges if $T(f)$ and $T(g)$ are finite. The coefficients a_i are given in terms of derivatives of $T(h)$ with respect to the variable t and therefore involve the difference $g(x) - f(x)$. For instance, the coefficient a_1 is found as

$$a_1 = \int u(x) \cdot [g(x) - f(x)]\, dx$$

or, equivalently

$$a_1 = \int \Psi(x)\, g(x)\, dx.$$

Before proceeding, we note that the closer h is to f, the smaller are the high-order terms. Thus, when g and f are close together, the expansion for $t = 1$.

$$T(g) = T(f) + \int \Psi(x)\, g(x)\, dx + \frac{1}{2} \int \int \varphi(x,y) g(x) g(x)\, dx\, dy + \cdots \qquad (2.28)$$

can be truncated to its first few terms. The terms are the so-called "derivatives". They involve functions $\Psi(x), \varphi(x,y), \cdots$ defined solely with the help of $T(f)$ and f. They cancel for equal distributions g and f.

It has not proved possible so far to state the conditions required to justify the above intuitive derivation. In this respect, von Mises only refers to previous works of Volterra, although they were found to be impracticable. Moreover, while a proper derivation is produced in

35

order to substantiate the expansion, many conditions appear which can hardly be related in an easy way to the functional $T(f)$ and to the distribution f and g (see Hill, 1977).

In order to provide some more insight into the mechanism underlying this Taylor-like expansion, we now present a derivation which supports the belief that most estimation procedures lead to differentiable statistical functions and thus satisfy (2.28). Because of the limited results we obtained we did not aim at establishing a perfect theory, leaving this objective to motivated analyzts.

For compactness, we propose a derivation with an ample use of operators and a particular one being the vector differential operator del, $\triangledown$, also called nabla. We shall recall briefly a few items to clarify the notation.

- The one-dimension (unlimited) Taylor's series. Let f be a real-valued function defined for an interval I in E^1. If $f \in C^\infty$ on I, namely if all derivatives $D^n f, n > 0$, do exist in each point of I, then for $x \in I$ and h sufficiently small we have:

$$f(x+h) = f(x) + \frac{h}{1!}Df(x) + \frac{h^2}{2!}D^2 f(x) + \frac{h^3}{3!}D^3 f(x) + \cdots$$

$$= [1 + \frac{h}{1!}D + \frac{h^2}{2!}D^2 + \frac{h^3}{3!}D^3 + \cdots] f(x)$$

$$= [exp(hD)] f(x)$$

- The multi-dimension (unlimited) Taylor's series. Under conditions similar to those mentioned above, we may write (underlining the vectors and the prime denoting the transposition)

$$f(x + h) = f(x) + \frac{1}{2!}h' \triangledown f(x) + \frac{1}{2!}(h' \triangledown)^2 f(x) + \ldots$$

$$= [1 + \frac{1}{1!}h' \triangledown + \frac{1}{2!}(h' \triangledown)^2 + \cdots] f(x)$$

$$= [exp(h' \triangledown)] f(x).$$

The first two m-th order differentials are easy to state in vector notation, then we must resort to specifying the components of h and ∇. We have

$$h' \, \nabla f(x) = \sum_i h_i \, D_i \, f(x)$$

$$(h' \, \nabla)^2 \, f(x) = [(h' \, \nabla)] \, f(x) \, (\nabla h) = h' \{\nabla [f(x)] \nabla'\} \, h$$

$$= \sum_i \sum_j h_i \, h_j \, D_{i,j} \, f(x)$$

$$(h' \, \nabla)^3 \, f(x) = \sum_i \sum_j \sum_k h_i \, h_j \, h_k \, D_{i,j,k} \, f(x).$$

- The general (unlimited) Taylor's series. Under mild regularity conditions (essentially the existence of the differentials), a scalar -, vector - or matrix - function f can be expanded in a Taylor's series in the vicinity of a scalar -, vector - or matrix-point according to the equation

$$f(x+h) = [\exp(h'\nabla)] \, f(x). \tag{2.29}$$

- The general limited Taylor's series. Whenever a truncated series is needed, it is appropriate to have at one's disposal an expression of the remainder term. Then we have for finite n

$$f(x+h) = [\sum_{k=0}^{n-1} \frac{1}{k!}(h' \, \nabla)^k + \frac{1}{n!}(h'\nabla)^n \exp(t \, h' \, \nabla)] \, f(x) \tag{2.30}$$

where

$$0 < t < 1.$$

Before going into the root of the derivation, we must state clearly the way in which we look at probability density functions. Accordingly with (2, 16), the unique purpose of a probability density function is to allocate a probability measure to a sample space or, in other words, to assign a probability to any subset of the sample space Ω. Thus defining a parameter ϑ by its value taken with respect to a probability density function f, i. e.

$$\vartheta = T(f),$$

37

is identical to defining it with respect to the probabilities

$$\mu_1, \mu_2, ..., \mu_N$$

taken on the elements of a partition of Ω. Technically we consider a finite partition for large cardinal number N such that all μ_i be small. They may for instance be upper-bounded according to

$$\mu_i < 1/\sqrt{N}.$$

These manipulations are needed to remain in the set-up of standard mathematical analysis; but this has drawbacks. The bounding of μ_i may not be admissible with some more-or-less discrete probability density functions, then a smoothing technique, such as (2.27), must be admissible. A more tricky aspect is that we assume ϑ to be uniquely defined by the set of infinitesimal probabilities μ_i, namely

$$\vartheta = T(\mu) \tag{2.31}$$

with

$$\mu = (\mu_1,...,\mu_M)'.$$

We observe that μ can be seen as the vector of the weights to be assigned to observations "centered" in each partition element. This is fully analogous to the definition of the Rieman-Stieltjes integral. It would be more correct to analyze the situation

$$\vartheta = \lim_{N \to \infty} T(\mu),.$$

but we will not and we avoid the problem of passing to an infinite partition by only considering consistent estimators. We have not said much about the manner in which we partition the sample space Ω. It seems safe, although possibly unnecessary, to impose a partition such that the element diameters decrease with increasing cardinal number N.

Given these preliminary items, the main derivation is trivial. We are interested in the relationships existing between the various values taken by an estimator $T(.)$ of some ϑ defined with respect to various probability distributions. Let the latter f and g and let the corresponding probability sets on a given sample space Ω be μ_0 and μ_1. Furthermore we assume that ϑ is defined with respect to the probability function f and estimated for g. Thus we have according to (2.31)

$$\vartheta = T(\mu_0) \tag{2.32}$$

and

$$\hat{\vartheta} = T(\mu_1). \tag{2.33}$$

We have already noted that (2.32) implies consistency of the estimator, but we now further restrict the class of $T(.)$ we investigate. Rather than seeing μ as a set of probabilities, it will be associated to a set of weights; hence the constraint (2.17) is not relevant any more. However some "homogeneity" in the weights is implied; namely for any (positive) constant c, the functional $T(.)$ must satisfy

$$T(\mu) = T(c\,\mu). \tag{2.34}$$

We introduce (2.33) and (2.32) into (2.29) and obtain

$$T(\mu_1) = \exp[(\mu_1-\mu_0)'\nabla]\, T(\mu_0)$$

$$= \exp(\mu'_1\,\nabla)\,\exp(-\mu'_0\,\nabla)\,T(\mu_0) \tag{2.35}$$

A special situation results from the substitution

$$\mu_0 = \mu,\ \mu_1 = 2\mu$$

which yields

$$T(2\mu) = \exp\,[(2-1)\mu'\,\nabla]\,T(\mu)$$

or in virtue of (2.34)

$$\exp(\mu'\,\nabla)\,T(\mu) = T(\mu); \tag{2,36}$$

this may also be written, for $k > 0$, as

$$(\mu'\,\nabla)^k\,T(\mu) = 0$$

Finally we substitute (2.36) into (2.35) and the following basic relationship results

$$T(\mu_1) = \exp(\mu'_1 \nabla)\, T(\mu_0) \tag{2.37}$$

When (2.32) and (2.33) hold good, the expanded notation is appropriate, thus

$$\hat{\vartheta} = T(\mu_1)$$

$$= \vartheta$$

$$+ (1!)^{-1} \sum \mu_{i,1}(\partial/\partial \mu_{i,0})\, T(\mu_{i,0}, \ldots, \mu_{i,0}, \cdots) \tag{2.38}$$

$$+ (2!)^{-1} \sum \sum \mu_{i,1}\, \mu_{j,1}\, (\partial^2/\partial \mu_{i,0}\, \partial \mu_{j,0})\, T(\cdots)$$

$$+ (3!)^{-1} \sum \sum \sum \cdots$$

or, generally speaking, between two distributions f and g,

$$T(g) = T(f) + \int \Psi(x)g(x)dx + \frac{1}{2}\int\int \varphi(x,y)g(x)g(y)dx\,dy + \cdots$$

which is precisely von Mises's result (2.28).

The successive terms of this expansion happen to be Gateaux's derivatives; for instance, the first von Mises's derivative is the first order Gateaux's derivative of $T(f)$ with respect to g, namely

$$\int \Psi(x)g(x)dx = \lim_{t \to 0} \{T[(1-t)f + tg] - T(f)\}/t$$

as can be easily verified. Moreover this provides a convenient way to evaluate $\Psi(x)$; we have

$$\Psi(x_0) = \lim_{t \to 0} \{T[(1-t)f + t\,\delta(x-x_0)] - T(f)\}/t \tag{2.39}$$

where $\delta(x-x_0)$ stands for the Dirac distribution centered on x_0.

An extremely interesting situation is a comparison with its asymptotic value ϑ of an estimator $\hat{\vartheta}$ based on an empirical distribution. The asymptotic distribution will be f in (2.28) whereas the empirical density of probability will be defined with respect to (possibly equal) weights w_i, i. e.

$$g(x) = (\textstyle\sum w_i)^{-1} \sum w_i \, \delta(x - x_i).$$ (2.40)

This leads to

$$\hat{\vartheta} = T(g)$$

$$= \vartheta + \frac{1}{1! \sum w_i} \sum w_i \, \Psi(x_i)$$ (2.41)

$$+ \frac{1}{2! (\sum w_i)^2} \sum \sum w_i \, w_j \, \varphi(x_i, x_j)$$

$$+ \frac{1}{3! (\sum w_i)^3} \sum \sum \sum \cdots$$

where the "coefficients" $\Psi(x_i), \Psi(x_i, x_j), \ldots$ are defined with respect to the distribution f and, therefore, are independent of the empirical distribution g. Another formulation of (2.41) may be very suggestive; it displays a structure which was also present in (2.38), namely

$$\hat{\vartheta} = \vartheta + \Psi'v + \frac{1}{2} v'\varphi v + \frac{1}{6} \cdots$$ (2.42)

where v is the column vector

$$v = (w_1 / \textstyle\sum w_i, w_2 / \sum w_i, \cdots)';$$

the row vector Ψ' and the matrix φ correspond respectively with $\Psi(x_i)$ and $\varphi(x_i, x_j)$ in (2.41). Naturally, v is subject to the condition

$$\textstyle\sum v_i = 1$$

or, in vector notation,

$$1'v = 1$$ (2.43)

where the column vector 1 is

$$1 = (1, 1, \cdots)'.$$ (2.44)

41

In what follows we shall need a well-known matrix identity that we demonstrate hereinafter, because of its importance. Let A be a full-rank square matrix, b and c be column vectors and λ be a (small) scalar. Then we have

$$
\begin{aligned}
(A - \lambda\, bc')^{-1} &= A^{-1/2} \left[\qquad\qquad I - \lambda A^{-1/2}\, bc\, 'A^{-1/2} \qquad\qquad\qquad \right]^{-1} A^{-1/2}\\
&= A^{-1/2} \left[\qquad\qquad\quad I \qquad\qquad\qquad\qquad\qquad \right] A^{-1/2}\\
&+ A^{-1/2} \left[\lambda\ A^{-1/2}\, b\left(\qquad\quad I \qquad\qquad\qquad \right)c\,'A^{-1/2} \right]\ A^{-1/2}\\
&+ A^{-1/2} \left[\lambda^2 A^{-1/2}\, b\left(c\,'\ A^{-1/2} \qquad\qquad A^{-1/2}b \right)c\,'A^{-1/2} \right]\ A^{-1/2}\\
&+ A^{-1/2} \left[\lambda^3 A^{-1/2}\, b\left(c\,'\ A^{-1/2}\, A^{-1/2}\, bc\,'A^{-1/2}\, A^{-1/2}b \right)c\,'A^{-1/2} \right]\ A^{-1/2}\\
&+ A^{-1/2} \left[\lambda^n A^{-1/2}\, b\left(\qquad\qquad \cdots \qquad\qquad \right)c\,'A^{-1/2} \right]\ A^{-1/2}\\
&= A^{-1/2} \left[I + \frac{\lambda}{1-\lambda c'\ A^{-1}b}\ A^{-1/2}bc'\ A^{-1/2} \right] A^{-1/2}\\
&= A^{-1} + \frac{\lambda A^{-1}bc'\ A^{-1}}{1 - \lambda c'\ A^{-1}b}\ .
\end{aligned}
$$

$$(2.45)$$

This derivation holds as long as

$$|\lambda\, c'\ A^{-1}\, b\,| < 1$$

but it is easy to verify that (2.45) is valid for any value of λ that does not lead to singularity.

Inspection of (2.2) might induce us to evaluate $\Psi, \varphi, \cdots$ by differentiating $\hat{\vartheta}$ with respect to the components of v. This must be done with care for v is constrained by (2.43). It seems more appropriate to proceed with respect to the w-components since they are not constrained. Indeed, we have

$$\frac{\partial \hat{\vartheta}}{\partial w_j} = \sum_i \left(\frac{\partial v_i}{\partial w_j} \right) \frac{\partial \hat{\vartheta}}{\partial v_i}$$

or

$$\frac{\partial \hat{\vartheta}}{\partial w} = \frac{1}{1'w}(I - 1v') \frac{\partial \hat{\vartheta}}{\partial v} \qquad\qquad (2.46)$$

where I stands for the identity matrix. What we are looking for is an inverse relationship of (2.46); but there exists an infinite number of such relations. So we constrain the problem by a further condition namely (2.34) indicating that $\hat{\vartheta}$ is "homogeneous" in w. (2.46) can be written with respect to a scalar λ as follows

$$\frac{\partial\hat{\vartheta}}{\partial w} = \frac{1}{1'w} \lim_{\lambda\to 1} (I - \lambda 1 v') \frac{\partial\hat{\vartheta}}{\partial v}$$

hence

$$\frac{\partial\hat{\vartheta}}{\partial v} = (1'w)\lim_{\lambda\to 1} (I - \lambda 1 v')^{-1} \frac{\partial\hat{\vartheta}}{\partial w}$$

$$= (1'w)\lim_{\lambda\to 1} (I + \frac{\lambda}{1-\lambda v' 1} 1 v') \frac{\partial\hat{\vartheta}}{\partial w}$$

in virtue of (2.45) and, on decomposing this and taking into account the homogeneity, we obtain

$$\frac{\partial\hat{\vartheta}}{\partial v} = (1'w)\frac{\partial\hat{\vartheta}}{\partial w} + \lim_{\lambda\to 1} \frac{\lambda}{1-\lambda v' 1} (1 w') \frac{\partial\hat{\vartheta}}{\partial w}$$

$$= (1'w)\frac{\partial\hat{\vartheta}}{\partial w} + \lim_{\lambda\to 1} \frac{\lambda}{1-\lambda v' 1} (w' \frac{\partial\hat{\vartheta}}{\partial w}) 1$$

$$= (1'w)\frac{\partial\hat{\vartheta}}{\partial w}.$$

The second term of the last but one line is identically zero whatever the value of λ. Hence

$$\frac{\partial\hat{\vartheta}}{\partial v} = (1'w)\frac{\partial\hat{\vartheta}}{\partial w}$$

or, written in terms of components:

$$\frac{\partial\hat{\vartheta}}{\partial v_i} = (\sum w_i) \frac{\partial\hat{\vartheta}}{\partial w_i}. \tag{2.47}$$

Considering that (2.41-42) is rapidly converging, which is a point to be further discussed, we can truncate the series at the level of its first order term and we obtain

$$\Psi = \partial\hat{\vartheta}/\partial v$$

under the constraint (2.43), or

$$\Psi(x_i) = \left(\sum w_i\right) \frac{\partial \hat{\vartheta}}{\partial w_i} . \qquad (2.48)$$

We turn now to consider the question of truncating the infinite series (2.41) or, rather, (2.37). We may see that, in some way, we should have used (2.30) rather than (2.29) to obtain (2.35) but then we would have missed a few important findings. The general limited Taylor's series (2.30) indicates that the series (2.41) can be truncated at any level provided a correction to the higher order term is applied (a single term is modified); the correction consists in evaluating the derivatives of (2.38) with respect to a probability density intermediate between f and g (rather than with respect to f). Hence whenever f and g are "close" together, truncation is allowed; but what means close is a little intuitive in this context. Let us say that whenever $T(g)$ is not very sensitive to perturbations of g (e. g., smoothing discrete distributions by (2.27) or moderately shifting the location of observations), then g is representative to a degree sufficient to suit $T(.)$ and should be "close" to the appropriate f-density of probability.

We now illustrate on a simple example the material we have presented in this section. We could have used the test case of §2.2 either, in only taking into account the type B observations, or in introducing conditional probabilities between types A and B. Both methods are too involved.

We work out the expansions of the weighted mean and variance estimated from a finite-size one-dimension sample. We need (and will need later) a proper method of normalizing the weights. It seems natural to introduce an average weight $\bar{w}$ so defined as if it were any weighted average. For any variate z_i, the arithmetic mean is

$$Ave(z_i) = \frac{1}{\sum w_i} \sum w_i z_i ,$$

we define $\bar{w}$ as the $Ave(w_i)$, e. g.

$$\bar{w} = \frac{1}{\sum w_i} \sum w_i^2 . \qquad (2.49)$$

In obvious notation, the expansions we are interested in are those relative to m and s^2 which are given by

$$m = \frac{1}{\sum w_i} \sum w_i\, x_i \qquad i=1,\dots,n$$

$$s^2 = \frac{1}{\sum w_i - \overline{w}} \sum w_i (x_i - m)^2 \tag{2.50}$$

The denominator in the s^2-expression stands for the ordinary $(n-1)$; it will be one of the outcomes of the jackknife theory (see 4.31).

With regard to the mean, the expansion (2.41) is limited to its first two terms. It is

$$m = \mu + \frac{1}{\sum w_i} \sum w_i (x_i - \mu) \ . \tag{2.51}$$

A few algebraic manipulations lead to the infinite expansion of the variance estimator. We have

$$s^2 = \frac{1}{\sum w_i - \overline{w}} \sum w_i\, [(x_i - \mu) - \frac{1}{\sum w_j} \sum w_j\, (x_j - \mu)]^2$$

$$= \frac{1}{\sum w_i - \overline{w}} [\sum w_i\, (x_i - \mu)^2 - \frac{1}{\sum w_j} \sum\sum w_i w_j (x_i - \mu)(x_j - \mu)]$$

$$= [\sum_{k=0}^{\infty} (\sum w_i^2)^k / (\sum w_i)^{2k+1}] \tag{2.52}$$

$$\cdot \ \{(\sum w_i)\sigma^2 + \sum w_i\, [(x_i - \mu)^2 - \sigma^2] - \frac{1}{\sum w_j} \sum\sum w_i w_j (x_i - \mu)(x_j - \mu)\}$$

$$= \sigma^2 + \sum_{n=1}^{\infty} \frac{c_n}{n!(\sum w_i)^n}$$

where the factors c_n have two different structures depending on whether they are odd or even. Both structures comprise a product of n weights w_i; they are consistent with (2.36) :

$$c_{2k} = (2k)!(\sum w_i^2)^{k-1} [(\sum w_i^2)\sigma^2 - \sum w_i w_j\, (x_i - \mu)\,(x_j - \mu)],$$

$$c_{2k+1} = (2k+1)!\, (\sum w_i^2)^k \sum w_i\, [(x_i - \mu)^2 - \sigma^2] \ .$$

45

The expression for $\Psi(x)$ can be obtained through identification in (2.41), via its Gateaux's derivative expression (2.39) as well as by its first order approximation (2.48). We shall now compare these methods for the mean and variance estimators.

The identifications of (2.51) and (2.52) with (2.41) yield respectively, for the mean,

$$\Psi(x) = x - \mu$$

and, for the variance,

$$\Psi(x) = (x-\mu)^2 - \sigma^2 .$$

To use the Gateaux's derivative approach we must consider
- a distribution f. This can be any distribution deemed reasonable in view of the investigated estimator. In this case, we assume it has a mean μ and a variance σ^2. Other features do not matter:
- a Dirac distribution centered in x_0, $\delta(x-x_0)$.
- an intermediate distribution for small t > 0,
   $h = (1-t)f + t \; \delta(x-x_0)$.

Then (2.39) can usually also be written as

$$\Psi(x_0) = (d/dt) \; T \; [h(t=0)]$$

For the mean estimator, we obtain

$$m(h) = (1-t)\mu + tx_0$$

and

$$\Psi(x_0) = x_0 - \mu .$$

The evaluation for the variance estimator is slightly more complex

$$s^2(h) = [(1-t)(\mu^2+\sigma^2) + tx_0^2] - [m(h)]^2$$

$$= \sigma^2 + [(x_0-\mu)^2 - \sigma^2] - t^2(x_0-\mu)^2$$

and

$$\Psi(x_0) = (x_0-\mu)^2 - \sigma^2 .$$

We turn our attention to the approximation (2.48) which, and this must be stressed, is exceptional owing to the fact that no prior knowledge of any underlying distribution f is required. We have, for the mean estimator,

$$\Psi(x_i) = x_i-m$$

and, for the variance,

$$\Psi(x_i) = \frac{1}{1-\overline{w}/(\sum w_i)}\{(x_i-m)^2 - s^2\,[1-(2w_j-\overline{w})/(\sum w_i)]\} .$$

In conclusion of these $\Psi(x)$ -evaluations, it is clear the identification with (2.41) and the Gateaux's derivative method (2.39) yield the function $\Psi(x)$ for any x, provided assumptions are made on the parent distributions $f(x)$. The first order approximation (2.48) gives the same results when the sample is representative of the parent distribution, but $\Psi(x)$ is defined only at the observation coordinates; the parent distribution does not have to be assumed. These are general findings which make us believe that assumptions on the parent distributions should be avoided whenever they introduce arbitrariness. In many cases those assumptions are not required; assumptions are only used to conceal a lack of knowledge (or, even, expertise).

Chapter 3

Robustness, breakdown point and influence function

3.1. Definition of robustness

In this section, we shall restate the viewpoint of Hampel (1968 - 1971) in different words. Eventually, at the end of § 3.3, we will be in a position to give a more convenient delimitation for the concept of robustness. In this work, we keep to Hampel's viewpoints although they are more and more frequently open to fundamental criticisms. Several attempts have been made to define robustness in more manageable terms, but no serious breakthrough has been achieved thus far. We feel that some new views will gradually gain credit and that it might be in the line proposed by Rieder (1980).

Let us consider a set of observations $\{x_1, \cdots, x_n\}$ drawn from some distribution $f(x)$; this set will be used to estimate some parameter ϑ, let $\hat{\vartheta}_n$ be the estimate. The sampling distribution of this estimate is noted as $\varphi(\hat{\vartheta}_n, f)$ and depends upon $f(x)$. Usually however we do not know $f(x)$ and only have a more or less valid model, say $g(x)$. Roughly speacking $\hat{\vartheta}_n$ is robust if it scarcely depends upon the difference between $f(x)$ and g (x) , i.e. we expect $\varphi(\hat{\vartheta}_n, f)$ and $\varphi(\hat{\vartheta}_n, g)$ to be close together.

More precisely, $\hat{\vartheta}_n$ is said to be robust with respect to distribution g (and to f) if

$$d(f, g) < \eta ==> d[\varphi(\hat{\vartheta}_n, f), \varphi(\hat{\vartheta}_n, g)] < \varepsilon \tag{3.1}$$

for small positive ε and η and for $d(.,.)$ from (2.19) .

Hampel (1971) also defines the ∏ robustness in order to relate estimators based on different sample sizes despite some lack of

48

independency between the samples.

We have to say that this definition is not quite realistic. The condition

$$d(f,g) < \eta$$

covers any distribution $f(x)$ in the η-neighborhood of $g(x)$ notwithstanding the fact that some $f(x)$, close to $g(x)$, can be not acceptable for extraneous reasons. Thus, it sometimes appears that a non-robust estimator is considered robust for some subset of distributions $f(x)$, particularly when $f(x)$ is restricted to be a member of some parametric family. For instance the ordinary arithmetic mean is not a robust estimator as we will see immediately but (at least technically) could be robust in some very constrained sense, such as in working out the mean value of the cosine of some random angle. Let us now show that the mean is non-robust.

Assume we intend to estimate the mean m of a size n sample $\{x_1, \cdots, x_n\}$ by

$$m = \frac{1}{n}\sum x_i.$$

The distribution model does not seem to be of major importance and it may be supposed to be Gaussian, namely the observations are supposed to be drawn from the one-dimensional distribution

$$g(x) = N(\mu, \sigma^2),$$

and thus we expect m to be distributed according to

$$\varphi(m,g) = N(\mu, \sigma^2/n).$$

These are ordinary assumptions, however we might have been unfortunate in selecting our model. In fact any model similar to $g(x)$ could have been responsible for the drawing of the sample at hand. Assume for instance that the real underlying distribution is the contaminated normal

49

$$f(x) = (1 - \eta) N(\mu, \sigma^2) + \eta N(a, \sigma^2)$$

with

$$\eta \; small$$

and

$$(a - \mu)^2 > n\sigma^2.$$

In such a situation the sample is very likely drawn from $g(x)$ and the mean is properly estimated but, from time to time, an observation will be drawn from the contaminating distribution with, as a result, a significant offset of the estimated mean. In fact the true distribution of the mean is the mixture of normal distributions

$$\varphi(m, f) = \sum_{l=0}^{n} \binom{n}{l} \eta^l (1-\eta)^{n-l} N(\mu + l(a-\mu)/n, \sigma^2/n);$$

it present a number of modes spaced at regular intervals between μ and a, although the global shape is very similar to the $\varphi(m, g)$-shape.

Let us now evaluate the Prokhorov distances between these distributions. We will not devote too much attention to exact estimations but will rather place emphasis on the main idea.

To estimate the distance between f and g, $d(f, g)$, we first decide what scale factor σ we have to introduce into definition (2.18) ; it will of course be the standard deviation of the involved normal distributions ($\sigma = \sigma$). We find the optimal set ω_0 directly from (2.22) ; for $a > \mu$, it is approximately given by

$$\omega_0 = [(a + \mu + \sigma)/2 - \sigma^2 \ln \eta / (a-\mu), \infty].$$

Therefore we obtain for the terms of (2.20)

$$F(\omega_0) \approx \eta,$$

$$G(\omega_0^\delta) \approx 0$$

and, by the use of (2.19),

$$d\,(f,g) \approx \eta.$$

In the same way, we evaluate the distance between the two distributions of estimator m under the two hypotheses f and g for the distribution of the sample. The following result,

$$d\,[\,\varphi\,(m,f),\varphi\,(m,g)\,] \approx 1 - (\,1-\eta\,)^n,$$

leads us to insert in (3.1)

$$\varepsilon \geq 1 - (1-\eta\,)^n > 1 - e^{-n\eta}$$

Hence, reconsidering the definition of robustness (3.1), we see that a small η does not necessarily imply a small ε : the arithmetic mean is not a robust estimator.

3.2. Definition of breakdown point

Sometimes ε in (3.1) remains small whatever is η as long as it remains inferior to some critical value $\eta^{\bullet}$. This value $\eta^{\bullet}$ is the so-called breakdown point as defined by Hampel (1971).

Precisely, we first select a small ε and define η according to

$$\eta^{\bullet} = \max \{ \eta : d(f,g) < \eta \Longrightarrow d\,[\,\varphi\,(\hat{\vartheta}_n,\,f),\,\varphi\,(\hat{\vartheta}_n,\,g)\,] < \varepsilon \,\}. \tag{3.2}$$

This first selection of ε causes $\eta^{\bullet}$ to possibly depend upon ε and this is not very satisfactory. However, in practice, it is observed that the above dependence is negligible in most cases.

In the way it has been presented, the breakdown point concept appears to be a curiosity. A more intuitive view relates to the possible offset of an estimator. Returning to the arithmetic mean estimator we have just been analyzing, we have demonstrated that small contaminations (small η) can yield any large offset of the mean (large $\varepsilon \leq 1$). In this case the breakdown point is $\eta^{\bullet} = 0$. The median of a one-dimensional distribution however exhibits an extreme robustness for it yields $\eta^{\bullet} =$

1/2. Up to half of the sample can be outlying without leading to a serious estimation error. An examination of the usefulness of the breakdown point has been reported by Hampel (1976) who placed special emphasis on a location estimation with one-dimensional samples.

We finally note that the above definition of robustness concurs with the minimax approach of Huber (1964) , particularly surveyed in two groups of lectures (1969, 1977c) . The idea already mentioned is to design an estimator to be the "best" with respect to the least favorable distribution of a distribution subset. Hence the estimator is not too sensitive to distributional features; Papantoni-Kazakos (1977) regards this as a stability factor.

3.3. The influence function

The breakdown point is a rough description of an estimator robustness and hence more detailed information can also be very convenient. Indeed, when an estimator is robust, it may be inferred that the influence of any single observation is insufficient to yield any significant offset. That influence is readily measured by the coefficients $\Psi(x_i)$ in (2.41) , i.e. in

$$\hat{\vartheta} = \vartheta + \frac{1}{\sum w_i} \sum w_i \Psi(x_i) + \cdots$$

Hampel (1974) has coined the term "influence curve" for the function $\Psi(x)$ because he was only considering one-dimensional samples. This is a very powerful tool to evaluate the robustness for one dimension - see Andrews et al. (1972) - as well as for several dimensions - for instance, Rey (1975b) - It can also be immediately generalized to situations where several parent distributions are concerned as in Rey (1975b) and Reid (1981), to applications on secondary estimators such as the variance (see Hampel et al, 1981) or very generally to most statistics as indicated by Rousseeuw and Ronchetti (1981). Basically it measures the sensitivity of $\hat{\vartheta}$ to each observation.

In conclusion, a prerequisite for robustness is that the influence function $\Psi(x)$ given by (2.39) or (2.48) be bounded. Mallows, in an experimental investigation (1975), fully supports the definition based on the first von Mises derivative, even for moderate size samples. His criteria involve a few "natural" requirements (consistency, being a statistic, everywhere defined, ...). Mallows also makes use of the expansion (2.41) to introduce a second-order influence function; it happens to be the integrand $\varphi(x,y)$ of the double sum and is defined as the influence function of the (first-order) influence function.

As an illustration we now derive the influence function for the estimator (2.13) as encountered in the analysis of § 2.2. Our first task consists in writing the mean-life estimator in terms of weighted observations. How to generalize to weights is not always obvious and may require some brain work. The main difficulty usually lies in maintaining some weight homogeneity. In our case, the denominator of

$$\hat{\vartheta} = \frac{1}{N_B - 1} \left(\sum_{i=1}^{N_A} t_{i,A} + \sum_{j=1}^{N_B} t_{j,B} \right)$$

can be readily transformed into the structure

$$\left(\sum_j w_{j,B} \right) - 1$$

which is correct for unit weights ($w_{j,B} = 1$), but incorrect for any other set of weights. The sensible way is to use the theory the estimator is originating from; the probability statement (2.7) then turns out to become:

$$Pr\{X_N | \vartheta\} \propto \prod_{i=1}^{N_A} [1 - G(t_{i,A})]^{\frac{w_{i,A}}{w}} \prod_{j=1}^{N_B} [g(t_{j,B})]^{\frac{w_{j,B}}{w}}$$

Hence, the weighted form of the estimator (2.13),

$$\hat{\vartheta} = \frac{1}{\left(\sum w_{j,B} \right) - \overline{w}} \left(\sum w_{i,A} t_{i,A} + \sum w_{j,B} t_{j,B} \right).$$

in which according to (2.49) the mean weight $\bar{w}$ is,

$$\bar{w} = (\textstyle\sum w_{i,A} + \sum w_{j,B})^{-1} (\sum w_{i,A}^2 + \sum w_{j,B}^2).$$

Direct application of (2.48) leads to the twofold influence function

$$\Psi(t_{i,A}) = (\textstyle\sum w_{j,B} - \bar{w})^{-1} [(\sum w_{i,A} + \sum w_{j,B}) \, t_{i,A} + (2\, w_{i,A} - \bar{w})\hat{\vartheta}],$$

$$\Psi(t_{j,B}) = (\textstyle\sum w_{j,B} - \bar{w})^{-1} [(\sum w_{i,A} + \sum w_{j,B}) \, t_{j,B} + (2\, w_{j,B} - \bar{w} - \sum w_{i,A} - \sum w_{j,B})\hat{\vartheta}]$$

whose values are listed in Table 5 of § 4.3.

Chapter 4
The jackknife method

4.1. Introduction

The so-called jackknife method has been introduced by Quenouille to reduce a bias in estimation and then progressively extended to obtain an estimate of variances. It is interesting in that it yields an improved estimator and enables the estimator to be assessed in an inexpensive way, that is with regard to the methodology although the computation, if required, is not necessarily inexpensive. By all standards the results thus obtained are impressive in most cases; but in a few, not very well defined, circumstances the results are either poor or ridiculous.

Unfortunately the scope of application of the jackknife method is not easy to demarcate. It may be seen, however, that it is applicable in the case of a proper regularity of the estimator with respect to the observations. And, according to Huber (1972) :

> It is hardly worthwhile to write down precise regularity conditions under which the jackknife has these useful properties, more work might be needed to check them than to devise more specific and better variance estimate.

As indicated in what follows, we disagree with the above viewpoint and support the use of the jackknife technique for all the estimators falling in the frame of § 2.4.

What about the jackknife method? We now present the technique with relatively few details and with a relatively straightforward notation used by Miller, because more involved considerations might obscure the main ideas and hence will be reserved for the next section.

The story starts in 1956, when Quenouille proposes to reduce the possible bias of statistical estimators through what appears to be a "mathematical trick". He observe that an estimator $\hat{\vartheta}$ based on a sample of size n can frequently have its bias expanded in terms of the sample size as follows

$$E\,(\,\hat{\vartheta} - \vartheta\,) = a_1 n^{-1} + a_2 n^{-2} + \cdots$$

This form is now compared with the corresponding one for the estimator $\hat{\vartheta}_i$ based on the sample of size $n-1$, i. e. the same sample but without the i-th observation. The expansion then becomes

$$E\,(\,\hat{\vartheta}_i - \vartheta\,) = a_1 (n-1)^{-1} + a_2 (n-1)^{-2} + \cdots$$

provided that the observations are independent. To reduce the bias, Quenouille proposes to consider the n variates

$$\tilde{\vartheta}_i = n\,\hat{\vartheta} - (\,n-1\,)\,\hat{\vartheta}_i$$

which have a similar bias expansion, except that the first order term cancels. It is

$$E\,(\,\tilde{\vartheta}_i - \vartheta\,) = -a_2 \,/\, [\,n\,(\,n-1\,)\,] + \cdots$$

Obviously, the same mathematical trick can be applied to the second leading term of the expansion, and so on. To avoid a loss of efficiency in the estimation, he suggests the following definition of an average estimator

$$\tilde{\vartheta} = (\,1/n\,) \sum_i \tilde{\vartheta}_i$$

$$= n\,\hat{\vartheta} - [\,(\,n-1\,)\,/\,n\,] \sum \hat{\vartheta}_i$$

For reasons that were not obvious at that time, the jackknife estimate $\tilde{\vartheta}$ frequently exhibits fairly good statistical properties. Tukey who will soon appear on the stage has also named $\hat{\vartheta}_i$, the pseudo-estimates, and $\tilde{\vartheta}_i$, the jackknife pseudo-values.

With regard to bias reduction it seems that it might be advantageous to modify the sample size n by deleting more than one observation, say h observations $(h \geq 1)$, and working out the pseudo-estimates with samples of size $(n-h)$. There are many ways of forming these subsamples and special consideration has been devoted to the three following schemes: first, deletion of ($h=1$) observation at a time; second, deletion of ($h>1$) consecutive observations and computation of $g=n/h$ pseudo-estimates; third, deletion of h observations in the $\binom{n}{n-h}$ possible different ways. The three schemes appear to be rather equivalent when the parameter h is moderate and the observations are strictly independent. When there is some serial correlation, the second scheme can be more reliable; the selection of h must be such that each group of h deleted observations be more or less independent of the other. The third scheme appears to be very valuable for theoretical derivations - see Sen (1977) - although it is of a limited practical interest. The above discussion is experimentally supported - see Miller (1974a) for references.

Before leaving the subject of bias reduction, generalizations of the jackknife method are worthy to mention. Instead of evaluating the pseudo-estimates from estimators based on different sample sizes, it is possible to take into account different estimators on the same samples. This is defended at length in the collection of papers by Gray and Schucany as well as in their book (1972).

The second significant progress in the history of the jackknife method takes place in 1958 when Tukey conjectures that the pseudo-values $\tilde{\vartheta}_i$ are representative of the incidences of each specific observation, therefore he proposes the jackknife variance estimate

$$\sigma^2(\hat{\vartheta}) = E[(\tilde{\vartheta}_i - \tilde{\vartheta})^2].$$

This conjecture will largely be supported in the sequel. We now simply fix our attention to the following most significant property of the jackknife method.

Bias reduction as well as variance estimation can be achieved without neither a detailed knowledge of the sample distribution nor an involved analysis of the estimation method. We only need a sample and an estimation definition.

57

The next section will demonstrate that the third central moment of $\hat{\vartheta}$ is also available.

The computation involved in the jackknife method may be very lengthy and, accordingly, one tends to apply it as much as possible by analytical means. One interesting case in point is due to Dempster (1966). Faced with difficulties in manipulating the deletion of one observation in his formulas, he states his developments with the help of a variable ε ranging from zero (no deletion) to one (complete deletion) and eventually confines himself to considering of the first and second order terms in the variable ε. This approach is also pursued in the case of the infinitesimal jackknife of Jaeckel (1972a) where the above variable ε is kept infinitesimal. It is of major importance for analytical derivations because, then, the differences between estimators can be stated in terms of derivatives. But, due to the lack of theoretical background, it has scarcely been published so far.

To conclude this introduction on the jackknife, method we would like to mention three papers which could introduce the reader to the jackknife. Mosteller (1971) in a naive presentation advocates its introduction into elementary courses of statistics at a high school level; Bissel and Ferguson (1975) place also some emphasis on the robust properties but say, after having demonstrated how beneficial the jackknife method can be:

> However the general warning still stands - the jackknife is sufficiently sharp to wound the unwary.

The very important review of Miller (1974) is possibly the most complete piece of information which tries to balance the advantages against the drawbacks of the method. Parr and Schucany (1980) propose a bibliography.

In the line of the jackknife but with very different objectives and theoretical set-up, we should mention the leave-one-out method of Lachenbruch (1968) which has only an intuitive justification, the work in survey sampling of Woodruff and Causey (1976) which is adapted to the survey context and the interesting paper of Gray, Schucany and

Watkins (1975). The last two papers are partly related to the infinitesimal jackknife.

The ordinary jackknife has been seen in the much more general setting of the bootstrap by Efron. But this is another story which we shall tell in the next chapter.

4.2. The jackknife advanced theory

In what follows we shall outline the main points of the derivations which justify the jackknife method. We will not discuss at length a technically important point namely that the estimators we are concerned with are assumed to be consistent. This attitude is supported by the observation that most estimators are solutions of functional equations (see Nakajima and Kozin, 1979).

Assume we have a recipe yielding an estimator $\hat{\vartheta}$, consistent with respect to some true value ϑ and based on a sample $\{ x_1,...,x_n \}$, each observation x_i appearing with the bounded non-negative weight w_i (the "true value" may be a distribution parameter of scalar, vectorial or functional nature). Then the estimator may be seen as a transformation of the set of observations and weights. Let it be written as

$$\hat{\vartheta} = T (x_1,...,x_n; w_1,...,w_n). \tag{4.1}$$

Hence, according to (2.41), it can frequently be expanded in the form

$$\hat{\vartheta} = \vartheta + (1/\sum w_i) \sum w_i \, \psi_i + \frac{1}{2}(1/\sum w_i)^2 \sum \sum w_i \, w_j \, \varphi_{ij} + \cdots \tag{4.2}$$

where the coefficients ψ_i and φ_{ij} are functions of the transformation T (.) and the sample $(x_1,...,x_n)$, but independent of the weights.

This model, limited to its first few terms, will be assumed throughout. We do not know whether its validity is strictly required but we know that it constitutes a sufficient condition for the validity of the jackknife method. Moreover, the jackknife fails in situations where the above model happens to be not verified.

Returning for a moment to § 2.4 in which estimators are regarded as functionals on distributions, we can say that the jackknife technique tends to infer information on $T(\mu)$ from (2.31) in spite of the fact that the only data set is μ_1 yielding $T(\mu_1)$ by (2.33). This attempt at inference may appear ridiculous, or at least illusory, when the representativity of the sample $\{x_1,...,x_n\}$ is an open question, under other conditions however it may be very sound. Mutatis mutandis, we may decompose μ into some eigen-components and assume that the dominant features are suitably represented in μ_1; in other words, the difference between μ and μ_1 is mostly with respect to the less significant eigen-components. This viewpoint stresses the relations between the sample representativity and the estimator appropriateness. Let us now return to our derivations.

Contrary to our introductory presentation, we consider arbitrary modifications of the set of weights to derive the pseudo-values. The original set of weights being

$$w = (w_1,..., w_n)',$$

we form the modified sets by changing the weights of the observations belonging to the i-th group of g groups. The number of groups, g, may be different from n and the groups may differ in size; however, the ordinary situation is n groups of size 1, that is to say each group is a single observation. The modified weight vectors are noted

$$w_i = (I + M_i)w \quad i = 1,...,g \tag{4.3}$$

where I stands for the identity matrix and M_i refers to the weight Modifications; in the ordinary situation M_i is diagonal with a non-zero constant on the rows corresponding to the i-th group observations and zeroes otherwise - Note that w_i is a vector (of weights) and should not be confused with the individual observation weights; the latter are scalars - Thus according to (4.1), the original estimator is

$$\hat{\vartheta} = T(X,w) \tag{4.4}$$

and we form the g pseudo-estimates by

$$\hat{\vartheta}_i = T(X, w_i) \tag{4.5}$$

The corresponding pseudo-values will be

$$\tilde{\vartheta}_i = [(w'_i 1)\hat{\vartheta}_i - (w'1)\hat{\vartheta}] / (w' M_i 1) \tag{4.6}$$

where the unit vector, 1, is given by (2.44). Eventually, we obtain the jackknife estimate by averaging the pseudo-values; thus:

$$\tilde{\vartheta} = [\sum (w' M_i 1)\tilde{\vartheta}_i] / \sum w' M_i 1 \tag{4.7}$$

Relating $\tilde{\vartheta}$ to $\hat{\vartheta}$ and ϑ is no easy matter in full generality. We will resort to a few simplifying tricks - Generalizing the notation of (4.2), we may write

$$\hat{\vartheta} = \exp\left(\frac{w'\ D}{w'\ 1}\right)\vartheta \tag{4.8}$$

similarly with (2.29); D stands for the differential operator. In particular, we have component-wise

$$\Psi_i = (D\ \vartheta)_i$$

and

$$\varphi_{ij} = (DD'\ \vartheta)_{ij}.$$

Furthermore, on account of the homogeneity in w of $\hat{\vartheta}$, the argument applied to obtain (2.36) leads to

$$\hat{\vartheta} = \exp\left(\frac{w'\ D}{w'\ 1}\right)\hat{\vartheta} \tag{4.9}$$

We note that this formalism conceals the role played by the observation values, x_i. In fact this role is completely taken into account by the operator D as it was by the coefficients $\Psi_i, \varphi_{ij}, \cdots$

In accordance with (4.5), the pseudo-estimates are

$$\hat{\vartheta}_i = \exp\left[\frac{w'\ (I + M_i)\ 'D}{w'\ (I + M_i)\ 1}\right]\hat{\vartheta}$$

and, inserted into (4.6), they yield the pseudo-values

$$\tilde{\mathfrak{J}} = \frac{1}{w'\,M'_i\,1} \sum_{l=0}^{\infty} \frac{1}{l!} \left\{ \frac{[\,w'\,(I + M_i\,)\,'D\,]^l}{[\,w'\,(I + M_i\,)'1\,]^{l-1}} - \frac{[\,w'\,D\,]^l}{[\,w'\,1\,]^{l-1}} \right\} \vartheta$$

$$= \vartheta + \frac{w'\,M_i\,D}{w'\,M'_i\,1} \vartheta$$

$$+ \frac{(w'\,1)\,w'\,M_i\,DD\,'\,M_i\,w + 2\,(\,w'\,1\,)\,w'DD\,'\,M_iw - (\,w'\,M_i1\,)\,w'DD\,'w}{2\,(\,w'\,1)\,[\,w'\,(I + M_i\,)'\,1\,]\,(\,w'\,M_i1)} \vartheta$$

$$+ \frac{1}{6} \cdots (D)^3 \cdots \vartheta + \cdots$$

Already the second order term in the expansion of $\tilde{\mathfrak{J}}_i$ happens to be relatively complex. Hence, and with a view to averaging the pseudo-values $\tilde{\mathfrak{J}}_i$ to obtain the jackknife estimate by (4.7), we constrain the modifying matrices as follows : we impose approximately equal denominators, i. e.

$$w'\,(I + M_i\,)\,'1 \approx w'\,(I + M_j\,)\,'1; \tag{4.10}$$

we impose that the summation of M_i yields a term cancellation, namely

$$(\,w'\,1\,)\,w'\,DD\,'\,(\textstyle\sum M_i\,)\,w = [\,w'\,(\textstyle\sum M_i)'1\,]\,w'\,DD\,'w,$$

and it turns out that the only solution is that this sum must be a multiple of the identity matrix

$$\textstyle\sum M_i = \lambda I. \tag{4.11}$$

This condition can be achieved in various ways but because M_i naturally has a diagonal structure, we further constrain the derivation by setting

$$M_i = t E_i \tag{4.12}$$

where t is a real number and E_i a diagonal matrix with ones on the rows corresponding with the observations to be modified in the i-th group. When the groups have a common size h, i. e.

$$h = 1'\,E_i\,1,$$

the factor λ in (4.11) is given by

$$\lambda = t\, g\, h\, /\, n \qquad\qquad (4.13)$$

where g is the number of groups and n stands for the sample size.

To conclude this derivation which yields the jackknife estimate, we will assume that on the average $(w'\, D)^l$ does not involve observations appearing in pairs, triples, quadruples, etc. - This is a very common feature (related to observation independency) we will soon discuss - Precisely, our expression "on the average" stands for an expectation taken on all possible samples x of size n. Thus, we assume for the m-th order differentials, while $m \geq 2$,

$$\text{Average } \{ D_{i,j,k,\dots} \} = 0, \text{ if not } i = j = k = \dots \qquad\qquad (4.14)$$

With regard to the previously introduced expansion of $\tilde{\mathfrak{J}}_i$, this corresponds to the assumption

$$DD' = \text{diag}\,(\,\partial^2 /\, \partial w_1^2,\, \partial^2 /\, \partial w_2^2 ,\, \cdots\,)\,;$$

which implies for instance

$$E_i\, DD'\, E_i = DD'\, E_i \qquad\qquad (4.15)$$

Combining items (4.10) to (4.15), we obtain the infinite expansion of the jackknife estimate,

$$\tilde{\mathfrak{J}} = \{\, \sum_{l=0}^{\infty} \frac{1}{l!} [\, \frac{(w'D)}{(w'1)} \,]^l \; \frac{2 + tl(l-1)}{2 + tl(l-1)h/n} + 0\,(\,t^2\, D^3\,)\, \}\vartheta \qquad\qquad (4.16)$$

A comparison between the two expansions (4.8) and (4.16) reveals the potentialities of the Quenouille's jackknife. We have

$$\hat{\vartheta} = \vartheta + (\,1/\sum w_i\,)\sum w_i\, \Psi_i + (\,1/\sum w_i\,)^2 [\, \sum_i \sum_{j>i} w_i\, w_j\, \varphi_{ij} + \sum_i \frac{1}{2}w_i^2\, \varphi_{ii}\,] + \cdots$$

and

$$\hat{\vartheta} = \vartheta + (\,1/\sum w_i\,)\sum w_i\, \Psi_i + (\,1/\sum w_i\,)^2 [\, \sum_i \sum_{j>i} w_i\, w_j\, \varphi_{ij} + \sum_i \frac{1}{2}\frac{1+t}{1+th/n}w_i^2\, \varphi_{ii}\,] + \cdots$$

The first differing item is the second order term which appears multiplied by factor $(1+t)$; this term is also the first which can introduce a bias decreasing with the sample size. Therefore its cancellation, obtained with $t = -1$, usually reduces the bias. We also note that a small t does not bring about any bias reduction.

With regard to bias reduction the following conclusions may be drawn for a large class of estimators :

-   If $\hat{\vartheta}$ can be expanded according to (4.2).
-   If the weights are independent of the observations x_i.
-   If the g groups have the same weights (see 4.10-13).
-   If each group is independent of the others (see 4.14-15).
    Then
-   The ordinary jackknife $(t = -1)$ possibly reduces the bias.
-   The jackknife estimate $\tilde{\vartheta}$ has the same asymptotic distribution as the original estimator $\hat{\vartheta}$, but for a translation resulting from the bias reduction.

This question of distribution we have just noted is the crucial link between the bias reduction feature of the jackknife and the variance estimation as proposed by Tukey. But let us first return to (4.14). We have assumed that "on the average" the estimator $\hat{\vartheta}$ does not depend on particularities relating to pairs, triples or, more generally, groups of observations; this precludes any correlation and, therefore, seems to force a condition of independency. Although independency is certainly sufficient, it is not necessary. In fact, correlation internal to each of the g groups may be tolerated because, in such conditions, (4.15) still holds true and so do the higher order similar expressions. Hence (4.14) is a little too constraining and a "block-structure" could also be accepted.

However, in any given case, the fact that (4.14) is satisfied on the average rather than strictly may be troublesome. To be more explicit we turn our attention to the case of the variance estimation we dealt with before. Estimating a weighted variance by (2.50), we obtained its expansion as (2.52) and, identifying it with (4.2) or (4.8), we may write

$$\varphi_{jj} = (DD')_{jj} \, \vartheta = 2 \, w_j^2 [\, \sigma_2 - (x_j - \mu)^2 \,]$$

$$\varphi_{jj} = (DD')_{jj} \, \vartheta = - (x_i - \mu) (x_j - \mu) , i \neq j.$$

The fact that on the average φ_{jj} cancels is here of no concern, although it results from the estimator unbiasedness and supports (2.49). Much more interesting are the terms $\varphi_{ij} (i \neq j)$; their average value is the population covariance. When the observations are independent, they do not contribute to any systematic effect, so that they are not important at all for bias reduction. The terms in $\varphi_{ij} (i \neq j)$, however, contribute to the sampling distribution of $\hat{\vartheta}$; but it turns out that they are exactly maintained in the jackknife estimate expansion. In conclusion, bias correction cannot compensate for a bias due to correlation and the sample distributions of $\hat{\vartheta}$ and $\tilde{\vartheta}$ are identical except for the little randomness which might be associated with bias reduction.

Let us now consider more closely the random contributions. Whenever (4.2) exhibits a fair convergence speed, the random contribution is essentially due to the first order term, e. g.

$$(1 / \textstyle\sum w_i) \sum w_i \, \Psi_i. \tag{4.17}$$

Furthermore, we have observed that it was exactly this term which occurred in the $\hat{\vartheta}$- and $\tilde{\vartheta}$- expansions and that, in the case of the jackknife estimate, this term resulted from a sum of components. Hence, we may think that somewhere in the jackknife procedure there is a possibility to grasp the random aspects. This is the case indeed and we will utilize it.

Consider the variate δ_i, which is constructed to be unbiased (see (4.7)),

$$\delta_i = \tilde{\vartheta}_i - \tilde{\vartheta} ; \tag{4.18}$$

with (4.12) and for $h=1$, it can be expanded into the series

$$\delta_i = [\, \Psi_i - (1 / \textstyle\sum w_j) \sum w_j \, \Psi_j \,] + \, \cdots$$

and thus is basically representative of the i-th group incidence on the estimator $\hat{\vartheta}$. The random contribution (4.17) appears to be approximately equal to the weighted mean of the values of δ_i and these values can be evaluated. We use this feature to derive the first two centered moments of the estimators $\hat{\vartheta}$ and $\tilde{\vartheta}$.

It seems useless to carry out the derivations on the full expansion (4.2) because then the results are so complicated that they cannot be interpreted in full generality. Eventually we would have to consider that expansions can be limited to their first order terms without introducing any large error. Hence we have preferred to truncate expansions right from the beginning of their presentation. In view of (4.8), (4.16) and (2.30), we assume the validity of the expansions

$$\hat{\vartheta} = \vartheta + (w'1)^{-1} \; \Psi \, 'w + bias \tag{4.19}$$

and

$$\tilde{\vartheta} = \vartheta + (w'1)^{-1} \; \Psi \, 'w \tag{4.20}$$

where Ψ stands for the row vector (or the matrix) of the first derivatives of the scalar $\hat{\vartheta}$ (or the vector $\tilde{\vartheta}$) with respect to the w-components taken in a suitable point. Then the variability of $\tilde{\vartheta}$ (for a given sample size n) is due to the variability of Ψ according to the sample - Remember that the weights cannot depend upon the observation values - In such circumstances, it is rather easy to proceed. But, since our results are more general than they would be under conditions (4.10 - 4.15), we will present the derivation independently.

First, the way in which we modify the sample (4.3) will be restricted to the general structure (4.12), namely

$$M_i = tE_i.$$

We further assume that the modifications are such that the g group appropriately overlap one another. This is expressed by

$$E_i \, E_j = 0, \; (i \neq j) \tag{4.21}$$

and

$$\sum_1^g E_i = I \tag{4.22}$$

Further restrictions to the way of forming the groups will be due to their respective weights and their respective independency - see (4.28).

The estimators of the centered moments result from a comparison between their definitions, namely, in virtue of (4.19) and (4.20) on the one hand,

$$\mu_r(\hat{\vartheta}) = \mu_r(\hat{\vartheta}) = (w'1)^{-r} \mu_r(\Psi'w) \tag{4.23}$$

and, on the other hand, the weighted arithmetic mean of δ_i

$$Ave(\delta_i^r) = [\sum (w' E_i 1)^r]^{-1} \sum [(w' E_i 1)^r \delta_i^r]. \tag{4.24}$$

We have here noted μ_r for the centered moment of order r and this implies that

$$\mu_2(.) = \sigma^2(.) = cov(.) ; \tag{4.25}$$

the second moment is a variance or a covariance matrix depending on whether $\hat{\vartheta}$ is a scalar or a vector. The same generalization could be made for higher order moments although it does not appear very useful. The shorthand "Ave" stands for an averaging on the observed items, further we will also need an expectation operator, which is denoted by "Expec" to avoid confusion. This operator refers to the (unknown) distribution of the sample and satisfies

$$Expec\,[Ave(\delta_i^r)] \;=\; Ave\,[Expec(\delta_i^r)] \;=\; [\mu_r(\delta_i)] \;. \tag{4.26}$$

To evaluate (4.24), we first derive δ_i from (4.6) and (4.18); it comes

$$\delta_i \;=\; \Psi'\,[(w' E_i 1)^{-1} E_i - (w'1)^{-1} I]\, w \;.$$

This is inserted into (4.24) and then manipulated with a view to finally taking the expectation of the result. To avoid cumbersome expressions we will substitute the following terms for:

$$a_i = -(w'E_i 1)/(w'1) \ .$$

$$b_i = \Psi'E_i w$$

and

$$c_i = w'E_i 1 = -a_i(w'1) \ .$$

We have

$$Ave\,(\delta_i^r) = [\sum_i (w'E_i 1)^r \ ]^{-1} \sum_i [\Psi' \, (E_i - \frac{wE_i 1}{w'1}I) \, w \,]^r$$

$$= [\sum_i c_i^r]^{-1} \sum_i [(1 - \frac{wE_i 1}{w'1}) \, \Psi'E_i w - \sum_{j \neq i} \frac{wE_i 1}{w'1} \, \Psi'E_j w \,]^r$$

$$= [\sum_i c_i^r]^{-1} \sum_i [(1+a_i)\, b_i \ + \ a_i \sum_{j \neq i} b_j \,]^r$$

$$= [\sum_i c_i^r]^{-1} \ \{\sum_i [(1+a_i)^r \, b_i^r \ + \ a_i^r \sum_{j \neq i} b_j^r] \ + \ (\cdots)\} \tag{4.27}$$

$$= [\sum_i c_i^r]^{-1} \ \{(\sum_i a_i^r) \, (\sum b_i^r)$$

$$+ \sum_i [(ra_i^{r-1} \ + \ \frac{r(r-1)}{2} \, a_i^{r-2} \ + \ \cdots) \, b_i^r] \ + \ (\cdots)\}$$

The term (...) involves cross-products of the form

$$b_i^{r-l} \, b_j^l \quad (l = 1, \ldots, r-1)$$

and they vanish when taking the expectation if and only if $r < 4$, in virtue of (4.21) and

$$Expec\,(b_i) = 0 \ ;$$

$$Expec\,[(\cdots)] = 0, r < 4 \ .$$

To decompose the last square bracket term of (4.27) we further assume independence between group-weights and group-moments, i. e.

$$Independency\ of\ (w'E_i 1)\ and\ \mu_r\ (\Psi'E_i w) \ . \tag{4.28}$$

Using (4.27) and (4.28), the left-hand side of (4.26) becomes

$$\text{Expec} \left[\text{Ave}\,(\delta_i^r) \right] = (-w'1)^{-r} \left[\sum_i \mu_r\,(b_i) \right] \left[1 + (\sum a_i^r)^{-1}\,\text{Ave}\,(r a_i^{r-1} + \cdots) \right] .$$

We now consider this expression for two special values of r, namely $r=2$ and $r=3$, while taking into account (4.22), e. g.

$$\sum_i a_i = -1 .$$

For the second order moment we obtain

$$\text{Expec}\,[\text{Ave}\,(\delta_i^2)] = \{ [\sum_i (w'E_i1)^2]^{-1} - (w'1)^{-2} \} \sum_i \mu_2\,(\Psi'E_iw) . \qquad (4.29)$$

The third order moment estimator is found as

$$\text{Expec}\,[\text{Ave}\,(\delta_i^3)] = [2 - 3g + g^2]\,(w'1)^{-3} \sum_i \mu_3\,(\Psi'E_iw) .. \qquad (4.30)$$

which has been written under the assumption of approximately equal group weightings, $w'E_i1$. Noting that

$$\sum_i \mu_2\,(\Psi'E_iw) = \mu_2\,(\sum_i \Psi'E_iw) = \mu_2(\Psi'w) ,$$

(4.23) and (4.29) lead to the covariance estimator

$$\text{Cov}\,(\hat{\vartheta}) \approx [(w'1)/\bar{w} - 1]^{-1}\,\text{Ave}\,(\delta_i\,\delta'_i) \qquad (4.31)$$

where $\bar{w}$ is the average weight

$$\bar{w} = \sum (w'E_i1)^2/\,(w'1) . \qquad (4.32)$$

Similarly (4.23) and (4.30) combine to give the third order moment estimator

$$\mu_3\,(\hat{\vartheta}) \approx [(g-1)\,(g-2)]^{-1}\,\text{Ave}\,(\delta_i^3) . \qquad (4.33)$$

It is convenient to express the last-mentioned two results in terms of the derivatives of (4.1),

$$\hat{\vartheta} = T(x_1,...,x_n;\, w_1,...,w_n) ,$$

with respect to the weight components w_i. For the sake of simplicity, we restrict ourselves to the case of independent observations; therefore we may form n groups $(g=n, h=1)$. The pseudo-estimates may be evaluated for any value of t but, because the moments do not depend upon this arbitrary parameter, we may as well consider an infinitesimal parameter t. Then the pseudo-estimates are

$$\tilde{\vartheta}_i = \hat{\vartheta} + t\, w_i\, \frac{\partial T}{\partial w_i}$$

The corresponding pseudo-values, from (4.6), are

$$\tilde{\vartheta}_i = \hat{\vartheta} + (\textstyle\sum w_j)\, \frac{\partial T}{\partial w_i}$$

and, as expected seeing (4.9) or (4.16), the jackknife estimate from (4.7) is equal to the original estimator

$$\tilde{\vartheta} = \hat{\vartheta} \ .$$

The variates δ_i defined by (4.18) are proportional to the partial derivatives, and happen to be the influence function values from (2.48),

$$\delta_i = (\textstyle\sum w_j)\, \frac{\partial T}{\partial w_i} = \Psi(x_i) \tag{4.35}$$

which lead to the following estimator of the variance-covariance matrix, according to (4.31) and (4.32).

$$Cov(\hat{\vartheta}) \approx \frac{(\sum w_j)}{(\sum w_j) - \overline{w}} \sum w_i^2 \, (\frac{\partial T}{\partial w_i})\, (\frac{\partial T}{\partial w_i})' \ . \tag{4.36}$$

where the average weight is

$$\overline{w} = \textstyle\sum w_i^2 / \sum w_i \ . \tag{4.37}$$

The third order centered moment is found from (4.33) as:

$$\mu_3(\hat{\vartheta}) \approx \frac{n^2}{(n-1)(n-2)} \sum w_i^3 \, (\frac{\partial T}{\partial w_i})^3 \ . \tag{4.38}$$

This is the end of the derivation of the jackknife theory. Let us further focus our attention to a few features of it.

For bias reduction as well as for centered moment evaluations, we had to assume independency between the g groups - see (4.14) and (4.28) - and this assumption may be open to question. However it is often possible to cope with the condition by selecting appropriate group sizes even when the individual observations are not independent. Moreover, the fact that averages have been used to evaluate the moments rather than the expectations favors the formation of as many groups as possible. Thus a compromise between group independency and a large number of groups might be needed. This consideration is opposite to Sharot's viewpoint (1976a), who assumes strict independency.

While computing the jackknife estimate, it is wise to check the range of the variates $(\tilde{\vartheta}_i - \tilde{\vartheta})$ in order to assess any presence of outliers, because they would impair the variance estimate. When robust estimators are jackknifed, these variates are bounded due to the limited incidence of each observation on the estimator $\hat{\vartheta}$. Hence the moment estimates are the more reliable, the more robust the estimator is.

In his 1972 memorandum, Jaeckel has proposed to modify the jackknife procedure in order to obtain bias-reduction even with the infinitesimal version. His proposal involves the second partial derivatives of T with respect to w_i and w_j and is, in fact, as much as possible an analytical duplication of the computational treatment with $(t = -1)$. We hesitate to recommend his derivation because other appropriate treatments could as well be advanced to mimic the full deletion of an observation. With a related standpoint the paper of Sharot (1976b) is noteworthy, his comparison of several types of jackknife variance estimators leads him to conclude that:

> On the basis of the Monte Carlo studies, it would appear that the desired gain in precision... is often achieved. Alternative estimators designed for a particular application may, not surprisingly, do better still. The infinitesimal jackknife is seen to yield just such an estimator in many cases.

We regret that the infinitesimal jackknife is so little known. It only appears in one page of Miller's review (1974), and there the reference to Jaeckel is of no great help.

To conclude this section on the jackknife theory, we point out that Thorburn (1976) enumerates the conditions to be satisfied by the transformation τ. He eventually states that the Taylor-like expansion we have assumed is required - see page 309, last but one equation. Nevertheless, we conjecture that the scope of application may be wider.

4.3. Case study

In what follows we are concerned with a numerical treatment of the jackknife method rather than with the analytical development we have mainly dealt with. The data set has been presented at § 2.2 and is summarized at Table 3; the mean-life estimator is given by (2.13) and we have discussed its influence function in § 3.3.

With as few references to the theory as possible, we present the ordinary jackknife method ($t = -1$) and the infinitesimal method (very small t), both methods seen in the equal-weight situation.

The (2.13) mean-life estimator

$$\hat{\vartheta} = S_T / (N_B - 1)$$

has a weighted form given by

$$\hat{\vartheta} = (\textstyle\sum w_j - \overline{w})^{-1} (\textstyle\sum w_i\, t_i)$$

where

$$\overline{w} = (\textstyle\sum w_i^2) / (\textstyle\sum w_i),$$

as indicated in § 3.3. It will be noted that the notation is here slightly condensed. We systematically use the index i for all observations whereas the index j only relates to the type-B observations. Thus

$$j = 1,\ldots, N_B$$

and

$$i = 1,\ldots, N = N_A + N_B.$$

The ordinary jackknife method requires the evaluation of the N pseudo-estimates $\hat{\vartheta}_i$ obtained with the same sample where successively each observation has been deleted. Thus, the pseudo-estimates are based on samples of size $(N-1)$. They are listed in the third column of Table 5. The fourth column gives the corresponding pseudo-values defined by (4.6) or

$$\tilde{\vartheta}_i = n\hat{\vartheta} - (n-1)\,\hat{\vartheta}_i.$$

Jackknifing the
Table 3 Life-time Data Set

Time	Type	Ordinary jackknife $\hat{\theta}_i$	$\tilde{\theta}_i$	δ_i	Infinites. jackkn. $\tilde{\theta}_i$	δ_i
1	A	44.88	46.75	10.30	52.50	7.50
3	B	51.00	-39.00	-75.45	-28.13	-73.13
4	B	50.86	-37.00	-73.45	-26.25	-71.25
5	A	44.38	53.75	17.30	60.00	15.00
8	B	50.29	-29.00	-65.45	-18.75	-63.75
8	B	50.29	-29.00	-65.45	-18.75	-63.75
9	B	50.14	-27.00	-63.45	-16.88	-61.88
10	B	50.00	-25.00	-61.45	-15.00	-60.00
12	A	43.50	66.00	29.55	73.13	28.13
20	A	42.50	80.00	43.55	88.13	43.13
25	B	47.86	5.00	-31.45	13.12	-31.88
40	B	45.71	35.00	-1.45	41.25	-3.75
40	A	40.00	115.00	78.55	125.63	80.63
75	A	35.63	176.25	139.80	191.25	146.25
100	B	37.14	155.00	118.55	153.75	108.75

$\hat{\theta}=45$ $\tilde{\theta}=36.45$

$\psi(x_i)=\delta_i$

Table 5

Eventually we obtain the jackknife estimate by averaging the pseudo-values.

It is given by (4.7) or

$$\tilde{\vartheta} = \frac{1}{n}\sum \tilde{\vartheta}_i = n\hat{\vartheta} - \frac{n-1}{n}\sum \hat{\vartheta}_i \qquad (4.39)$$

73

and has here the value

$$\tilde{\vartheta} = 36.45$$

which is to be compared with the value of the original estimator

$$\hat{\vartheta} = 45$$

The difference between $\tilde{\vartheta}$ and $\hat{\vartheta}$ is the result of the bias reduction procedure.

It may be thought that the bias reduction is illusory however, considering estimators which should introduce different biases, it appears in Table 4 of § 2.2 that the jackknife estimates are much more consistent.

To obtain centered moment estimates, we first evaluate the variates δ_i from (4.18)

$$\delta_i = \tilde{\vartheta}_i - \tilde{\vartheta} ; \tag{4.40}$$

which are representative of the random contributions to the estimators $\hat{\vartheta}$ and $\tilde{\vartheta}$. They are listed in the fifth column of Table 5 and lead to the variance estimate by (4.31)

$$Var\,(\hat{\vartheta}) \approx Var\,(\tilde{\vartheta}) \approx \frac{1}{n\,(n-1)} \sum \delta_i^2 \tag{4.41}$$

which here takes the value

$$Var\,(\hat{\vartheta}) \approx Var\,(\tilde{\vartheta}) \approx 340.0$$

The third centered moment can be estimated in exactly the same way. It is given by (4.33) or

$$\mu_3\,(\hat{\vartheta}) \approx \mu_3\,(\tilde{\vartheta}) \approx \frac{1}{n\,(n-1)\,(n-2)} \sum \delta_i^3 . \tag{4.42}$$

This reveals a marked asymmetry of the sample distributions

$$\mu_3\,(\hat{\vartheta}) \approx \mu_3\,(\tilde{\vartheta}) \approx 1132.8.$$

The infinitesimal jackknife method can be presented in the same way as the previous ordinary jackknife method. However the theory

74

tells us a few facts which we must not ignore. First, no bias reduction takes place and the jackknife estimate is exactly the original estimator (the pseudo-values $\tilde{\vartheta}_i$ are listed in the last but one column of Table 5). Second, the variates δ_i given by (4.35) are influence function values, $\Psi(x_i)$; their analytical form has been derived at 3.3 and they are contained in the last column of Table 5. They yield the variance estimator (4.41) or

$$Var\ (\hat{\vartheta}) \approx 331.7$$

The third centered moment is estimated by (4.42), i. e.

$$\mu_3\ (\hat{\vartheta}) \approx 1204.5$$

The example we have presented relates to a scalar estimator. It could have been interesting to see whether multidimensional estimators can be analyzed with the same elegancy. The answer is positive as will appear in what follows but we would like to call the attention on a nice situation we reported (1978, pages 26-27) where part of the observations was missing. There, the standard covariance estimator was not even positive definite whereas the jackknife method produced a pleasant inexpensive substitute.

With regard to applications, they can be found in various domains; some are interesting because it has been observed that the jackknife apparently fails on correlated data and on order statistics. This is not so surprising to us. The experience with this technique is gradually developing, applications and further theoretical developments can be found in many Journals.

4.4. Comments

On tendency to normality. It is known that under regularity conditions and with large group-size h as well as finite group-number g, the jackknife estimate tends to have an Hotelling's T-distribution with $g-1$ degrees of freedom. To us, this appears accidental and mainly a consequence of the mathematical derivation. Expressing the estimator $\hat{\vartheta}$ in terms of the observations x_i as a power series expansion and truncating at a low order necessarily produces pseudo-variates $\hat{\vartheta}_i$ with normal distributions if their components are many and bounded (appropriately disguised in the regularity conditions). And the conclusion is reached that the jackknife estimate tends to normality. This situation has been avoided in this text by expanding in terms of weights rather than observations. It frequently occurs that the jackknife estimate is more or less normal but, then, the original estimator $\hat{\vartheta}$ was also subject to the application of the Lindeberg conditions (see Reeds, 1978).

On time-series analysis. Jackknifing can be applied under conditions of independency. When some stationarity in variances can be assumed and sample size permits, groups of relatively large sizes h must be assembled. All other conditions are generally fulfilled.

On order statistics. This presentation does not justify the utilization of the jackknife. In fact, the estimators are usually not continuous in the observation weights and the weights are not independent of the observation values.

On misclassification probabilities. Numerous methods are in use to classify, among several classes, an extraneous sample on the basis of prior information. This prior information frequently consists in a set of observations belonging to known classes. Then, the probability of misclassification can be considered as a function of these known observations. An estimator of the misclassification probability can be jackknifed in "discriminant analysis", where this estimator depends smoothly on observation importances; but application of the method with some other classification techniques, such as "nearest neighbor", is not justified.

On transformations. If the jackknife method is applicable to $\hat{\vartheta}$, it is also applicable to $\varphi = \varphi(\hat{\vartheta})$ inasmuch as $\varphi(.)$ is continuously once differentiable in the vicinity of $\hat{\vartheta}$. However, the transformation may be useful to set confidence domains when the distribution of φ is more easy to manipulate than the original distribution - see Cressie (1981).

On robustness. The jackknife method modifies an estimator and does not have the power to render robust what was not robust. However it involves arithmetic means (4.24) which can be replaced by trimmed means (7.13) as suggested by Hinkley and Wang (1980). Such a substitution should not modify the limit laws investigated by Leprêtre (1979).

Chapter 5
Bootstrap methods,
sampling distributions

5.1. Bootstrap methods

In two similar papers, Efron (1979a, 1979b) has advanced views on the very old problem of estimation. Do we need distribution models? Should we rely on distribution models? His answer is a definite challenge and could give a new sense to the fundamental concepts. Let the sample speak for itself and do not interfere with extraneous assumptions. This is strictly consistent with our own views.

Efron's two papers are also provoking in the way they are presented - Note that the joined work of Diaconis and Efron (1983) is less provoking - They appear to be very simple reading because they clearly stress the main ideas; many technical details however are either absent or concealed behind (sometimes arbitrary) statements interlaced with conjectures (and printing errors in the equations). This mixture of competitive items leaves the reader puzzled and a little uncomfortable. These papers also indicate the overwhelming self-confidence of the author; this is reflected for instance in his acknowledgment section:

> I also wish to thank the many friends who suggested names more colorful than "Bootstrap", including... my personal favorite, the "shotgun" which, to paraphrase Tukey, "can blow the head off any problem if the statistician can stand the resulting mess.

Can we stand the mess? Most certainly, if no mess results and this looks like being the case.

Let us first recall what we have already observed before, i. e. that an estimator definition (a recipe or an algorithm) and a data set could yield a great deal of information on the sampling distribution of the estimator without any further modeling: the influence function (2.48), the bias reduction and centered moment estimations by the jackknife method do not require model assumptions. This is used by Efron who quite generally considers that the sample at hand can be regarded as the best representative of its own distribution and this is indeed true in that all the usual statistical inferences should be drawn from the empirical distribution. We have to be a little more precise in this matter.

Let us assume that we intend to estimate some parameter ϑ by the estimator $\hat{\vartheta}$ based on the sample $\{s_1, \ldots, x_n\}$ drawn from the unknown distribution $f(x)$. Then we have

$$\vartheta = T(f) \tag{5.1}$$

and

$$\hat{\vartheta} = T(g) \tag{5.2}$$

with, as in (2.40),

$$g(x) = \left(\sum w_i\right)^{-1} \sum w_i \, \delta(x - x_i) . \tag{5.3}$$

The parameter $\hat{\vartheta}$ has a sampling distribution which could be observed by repeatedly sampling from $f(x)$, the unknown distribution, if we had such a possibility; Efron proposes to sample from $g(x)$ rather than from $f(x)$. Then, a population of estimators $\{\vartheta_1^{\bullet} \cdot \vartheta_2^{\bullet} \cdots \}$ can be constructed and analyzed. Let them be formulated as:

$$\vartheta_j^{\bullet} = T(h_j) \tag{5.4}$$

where

$$h_j(x) = \left(\sum c_{ij} \, w_i\right)^{-1} \sum c_{ij} \, w_i \, \delta(x - x_i) . \tag{5.5}$$

The empirical distributions h_j are derived from $g(x)$. The same sample size n is achieved and the same observations are present, some of them once, others twice or three times, while a few of them are absent. The distributions $h_j(x)$ are obtained by a sampling with resubstitution from

79

$g(x)$; therefore the coefficients c_{ij} are integers. They satisfy

$$\sum_i c_{ij} = n \tag{5.6}$$

and are distributed according to the multinomial law

$$(p_1, \cdots, p_n)^n \tag{5.7}$$

or

$$Prob \ \{c_{1j}, \cdots, c_{nj}\} = (n!) \prod (c_{ij}!)^{-1} \ p_i^{c_{ij}} \tag{5.8}$$

with

$$p_i = 1/n \tag{5.9}$$

We note that Efron indicates the possibility of various other sampling schemes which generally are much less random than is the above sampling with resubstitution. For instance the ordinary jackknife pseudo-estimate $\hat{\vartheta}_i$ corresponds with

$$c_{ij} = 0, \ \text{if} \ i=j$$

$$= 1, \ \text{if} \ not.$$

The conjecture that the population of estimators $\{\vartheta_1^{\cdot\cdot}\vartheta_2^{\cdot}, \cdots \}$ is indeed distributed according to the sampling distribution of $\hat{\vartheta}$ by (5.2) will be (moderately) supported in the next section. For the time being we want to turn attention to a special feature. Many estimators are selected according to given properties of the data set; these properties can be essential and nevertheless be jeopardized by the sampling scheme. Then, obviously the bootstrap method cannot be beneficial. Such a case would occur when bootstrapping an estimator based on a balanced design. This is so evident that it may appear ridiculous to stress the point, but the situation could be less evident as it is in the following case.

For the last time we analyze the estimator (2.13) of mean-life applied to the data set reported in Table 3. The bootstrap method has been run in the standard way of sampling with resubstitution and a population of bootstrap estimators $\{\vartheta_1^*, ..., \vartheta_{1000}^*\}$ with size $n = 1000$ has been obtained. This population is displayed at Fig. 4 in the form of a stem-and-leaf display according to Tukey (1977). It might be observed that the median is approximately 44.2 and that the first and third quartiles are about 33.9 and 58.0 respectively. This is consistent with a moderately skew distribution with a scatter

$$\hat{\sigma} = 0.741 \; interquartiles = 17.86 \; .$$

and in close agreement with the jackknife result 18.44 by (4.40-41); however the right tail of the distribution is just incredible. Indeed the main conjecture of the bootstrap is invalidated in the present situation. What happened is that the use of (2.13) implies that a few type B observations have been made; nobody would try to estimate the mean life if only censored observations (possibly except one) were in the data set at hand. The bootstrap method ignores such a restriction. Running this computer experiment with a size 1000, we could by chance have generated a bootstrap sample with $N_B = 0$ or with $N_B = 1$; then a negative estimator or an infinite estimator ϑ_j^* would have resulted. The risk of that bootstrap collapse was 0.025. Needless to say that any blind use of the bootstrap can be harmful.

Commenting on the illustrative case we have presented, B. Efron wrote us in his letter dated August 6, 81:

In those cases where the bootstrap distribution has a long tail I often use a more robust estimate of standard deviation, for example one based on the interquartile range of the bootstrap distribution. The fact is that if indeed there is a real probability that N_B could be very small, then the bootstrap distribution is telling you about the resulting instability. Alternatively, you might wish to bootstrap in such a way that N_B is kept at its observed value, though I don't know exactly how to do that in the case at hand. (It would depend on the assumptions you are making relating the t_i to the labels A and B).

It will be noted that the proposal of keeping N_B at its observed value contradicts the issue defended in Efron (1981b) on the treatment of censored data - intermediate views are justified by Reid (1981) who discusses the information present in the censoring process and

```
   0     6 |
   1     8 | 8
   2    10 | 0
   3    12 | 2
   7    14 | 4445
  16    16 | 666667777
  30    18 | 88889999999999
  51    20 | 000000000000011111111
  84    22 | 2222222222222222222233333333333333
 117    24 | 444444444444444455555555555555555
 144    26 | 66666666666666667777777777777
 176    28 | 888888888888888999999999999999999
 211    30 | 00000000000000000011111111111111111
 253    32 | 22222222222222223333333333333333333333333
 304    34 | 444444444444444444444444445555555555555555555555555
 355    36 | 6666666666666666666666666677777777777777777777777
 403    38 | 88888888888888888888888889999999999999999999999
 452    40 | 0000000000000000000000001111111111111111111111111
 497    42 | 222222222222222222233333333333333333333333
 534    44 | 44444444444444444455555555555555555
 576    46 | 666666666666666666667777777777777777777
 613    48 | 8888888888888888899999999999999999999
 647    50 | 00000000000000000111111111111111
 680    52 | 2222222222222222222222233333333333
 712    54 | 444444444444444455555555555555
 750    56 | 6666666666666666666666677777777777777777
 783    58 | 8888888888888888888999999999999
 807    60 | 000000000000000111111111
 834    62 | 22222222222222223333333333333
 850    64 | 4444444555555555
 877    66 | 66666666666666677777777777777
 894    68 | 888888999999999
 914    70 | 00000000001111111111
 925    72 | 22223333333
 928    74 | 455
 933    76 | 77777
 937    78 | 8889
 951    80 | 00000111111111
 955    82 | 2222
 960    84 | 44455
 967    86 | 6667777
 969    88 | 88
 972    90 | 001
 973    92 | 3
 973    94 |
 977    96 | 6667
 977    98 |
 983   100 | 000111
 985   102 | 23
 985   104 |
 987   106 | 77
 989   108 | 88
 990   110 | 0
 991   112 | 3
 991   114 |
 992   116 | 6
 992   118 |
 994   120 | 01
 994   122 |
 994   124 |
 994   126 |
 994   128 |
 994   130 |
 995   132 | 3
 995   134 |
 995   136 |
 995   138 |
 997   140 | 01
 997   142 |
 997   144 |
 998   146 | 6
 998   148 |
 998   150 |
 998   152 |
 998   154 |
 998   156 |
 998   158 |
 998   160 |
 998   162 |
 998   164 |
 998   166 |
 998   168 |
 998   170 |
 998   172 |
 999   174 | 5
 999   176 |
 999   178 |
 999   180 |
 999   182 |
 999   184 |
 999   186 |
1000   188 | 9
1000   190 |
```

Fig.4.
Stem-and-leaf Display.
Bootstrap estimates obtained
for the life-time data set.

proposes to accordingly adjust the sampling scheme.

Evidently the bootstrap method must be applied with a little care, but this does not prevent its use when we see its outcomes as complementary to those produced by the more standard methods.

Before we examine the relationships existing between the estimator sample distribution and the distribution exhibited by the bootstrap method, we want to stress that the bootstrap failure we have reported is much more than accidental, it is intrinsic to the method while applied in certain contexts. For instance, in linear regression, the bootstrap sample has a small probability to belong to a subspace and have a singular design matrix; hence, generating quite a few bootstrap samples, there is some probability of meeting with a failure of the bootstrap method. In this regard, Freedman (1981) insists on the details of the bootstrap sampling scheme.

5.2. Sampling distribution of estimators

Not much can be said about the sampling distributions except in relatively constrained situations. The basic tool will be considering the distributions through their characteristic functions while the estimators are limited to their first order terms of their series expansions after the weights and this is much more elementary than the techniques applied by Singh (1981) and less asymptotic than the derivations of Bickel and Freedman (1981).

Since we encounter several distributions in what follows, we have introduced a systematic notation for the characteristic function. The characteristic function $\overset{o}{f}(u)$ of the variate x distributed according to $f(x)$ is defined as the expectation of $[\exp(i\,u'x)]$. An empty dot placed above the distribution script denotes the characteristic function and the vector u has the same dimensionality as the variate x; thus

$$\overset{o}{f}(u) = \int_{\Omega} e^{i\,u'x}\,f(x)dx\,,\tag{5.10}$$

the integration is performed over the sample space Ω and has to be understood in terms of the distribution theory as given in (2.16). We shall now establish a very general formula useful for our purposes.

Consider the following situation. A variate y (of any dimension) is obtained by premultiplying an n-dimensional variate x by a constant matrix A,

$$y = A x = \sum_j x_j a_j \tag{5.11}$$

with

$$x = (x_1, \cdots, x_n)'$$

and

$$A = (a_1, \ldots, a_n) \ .$$

The components x_j are distributed according to the multinomial law

$$(p_1, \cdots, p_n)^m \tag{5.12}$$

or

$$Prob \ \{x\} = (m!) \prod (x_j!)^{-1} p_j^{x_j}$$

with

$$\sum x_j = m$$

and

$$\sum p_j = 1 \ .$$

Then the characteristic function of y distributed according to $h(y)$ is derived as follows

$$\overset{\circ}{h}(u) = \int \exp \ (i \ u'y) \ h(y) \ dy$$

$$= \sum_x \exp \ (i \ u'Ax) \ Prob \ \{x\}$$

$$= \sum_x \exp{(i\, u'a_j x_j)}\; Prob\,\{x\} \tag{5.13}$$

$$= (\sum_{j=1}^n p_j\, e^{i\,u'a_j})^m \; .$$

The multidimensional characteristic function, treated in the ordinary way, leads to the distribution first moments. We have, for instance,

$$Mean\,(y) = -i(\partial/\partial u)\,\overset{o}{h}(0) \tag{5.14}$$

$$= m\,\sum_j p_j\, a_j \tag{5.14}$$

$$Cov\,(y) = -[(\partial^2/\partial u^2)\,\overset{o}{h}(0) + (Mean)\,(Mean)\,']$$

$$= m\,[\,\sum_j p_j\, a_j\, a'_j - (\sum_j p_j\, a_j)\,(\sum_j p_j\, a_j)'\,]\; . \tag{5.15}$$

The above formulation is appropriate to place most estimators in a common frame. First we associate a probability measure with the sample space Ω according to (2.32), namely

$$\vartheta = T(\mu_0)\; .$$

with

$$\mu_0 = (\mu_{1,0},\, \cdots\, ,\mu_{N,0})\; '\; .$$

a vector of infinitesimal probabilities. Then, by the use (2.33) and (2.38), we express the estimator $\hat{\vartheta}$ based on a sample of size m as

$$\hat{\vartheta} = T(\mu_1)$$

$$\approx \vartheta + \sum_j \mu_{j,1}\,(\partial/\partial\mu_{j,0})\, T(\mu_0) \tag{5.16}$$

where the N coefficients $(m\,\mu_{j,1})$ are distributed according to the multinomial law

$$(\mu_{1,0},\, \cdots\, ,\mu_{N,0})^m \; . \tag{5.17}$$

85

Eventually, comparing (5.16-17) with (5.11-12), we obtain for the sampling distribution of $\hat{\vartheta}$, i. e. for $g(\hat{\vartheta})$,

$$\overset{\circ}{g}(u) = e^{iu'\vartheta}\{\sum_{j=1}^{N} \mu_{j,0} \exp[i\,u'\,(\partial/\partial\mu_{j,0})\,T(\mu_0)/m]\}^m \ .$$

It is observed that, letting N tend to infinity, this expression becomes

$$\overset{\circ}{g}(u) = e^{iu'\vartheta}\{\int_{\Omega} \exp[i\,u'\Psi(x)/m]\,f(x)\,dx\}^m \tag{5.18}$$

where $f(x)$ stands for the density of probability of the variate x. Moreover, rather than investigating an equally weighted structure we could have used arbitrary weights. Then, writing (2.41) similarly to (5.16), we have

$$\hat{\vartheta} \approx \vartheta + \sum_{j} (m\mu_{j,0})\,(w_j/\sum_{k} w_k)\,(\partial/\partial\mu_{j,0})\,T(\mu_0) \ .$$

where $(m\mu_1)$ is distributed according to (5.17) as previously. Thus, taking into account (5.12-13), the characteristic function $\hat{\vartheta}$ has the form

$$\overset{\circ}{g}(u) = e^{iu'\vartheta}\{\int_{\Omega}\exp[i\,u'W\Psi(x)/(\sum w_k)]\,f(x)\,dx\}^m \tag{5.19}$$

Application of (5.14), under the constraint (2.36), leads to the unbiasedness of $\hat{\vartheta}$, namely

$$Mean\,(\hat{\vartheta}) = \vartheta \ . \tag{5.20}$$

Whereas (5.15) provides a covariance matrix

$$Cov\,(\hat{\vartheta}) = [\sum w_k^2/(\sum w_k)^2]\int \Psi(x)\,\Psi(x)'\,f(x)\,dx \ . \tag{5.21}$$

This formula closely resembles (4.31), in accordance with (4.35). However, we must recall that (5.20) and (5.21) have been obtained under the assumption of valid truncation of the infinite series (2.38) and (2.41).

Under the same hypothesis of valid truncation, it is useless to derive the distribution of the jackknife estimate $\hat{\vartheta}$ because, then, it equates the original estimators $\hat{\vartheta}$ and has the same properties (5.19-

21). Therefore we turn our attention directly to the bootstrap estimators ϑ_j^* according to (5.4-9).

Let us first observe that our interest bears on the conditional distribution of ϑ_j^*, being a conditionality with respect to a given sample distribution expressed by (5.3). Relaxing that conditionality would mean an access to the underlying $f(x)$ and we could then derive ϑ by (5.1); which is absurd. A second observation is that so far we have not devoted much attention to the way the weights are set; this depends very much upon the treated problem and we have only required the independence of w_i with respect to x_i. The dependence of the bootstrap estimators on the coefficients c_{ij} in (5.5) may also be of importance. In what follows, we avoid this difficulty by only considering the equal-weighting case.

Thus we are interested in the distribution of ϑ_j^* given $g(x)$; let it be denoted by $h(\vartheta_j^*|g)$. The following expressions are derived from (2.41) for a sample of size m. Whereas the original estimator $\hat{\vartheta}$ is approximately given by

$$\hat{\vartheta} = \vartheta + m^{-1} \sum_k \Psi(x_k) \, , \tag{5.22}$$

the bootstrap estimators obey the formula

$$\vartheta_k^* = \vartheta + m^{-1} \sum_k c_{kj} \, \Psi(x_k) \tag{5.23}$$

with the c_{kj} distributed according to the multinomial law

$$(m^{-1}, \cdots, m^{-1})^m \, .$$

in the ordinary bootstrap where sampling with resubstitution is performed. Then, according to (5.11-13), the characteristic function is

$$\overset{\circ}{h}(u\,|g) = e^{iu'\vartheta} \, \{m^{-1} \sum_k \exp \, [iu'\Psi(x_k)/m\,]\}^m \, . \tag{5.24}$$

This result leads us to support the Efron's conjecture that ϑ_j^* has the same distribution as $\hat{\vartheta}$ inasmuch as the term between braces of (5.24) is a finite approximation of the similar term between braces of (5.18).

To sum up, the bootstrap method is valid when approximations (5.22-23) can be justified. However it must be observed that (5.22) is more valid than (5.23), for $g(x)$ in (5.3) is closer to $f(x)$ in (5.1) than $h(x)$ by (5.5) can be. The net result is that moment estimations should be performed by the jackknife method rather than by the bootstrap method.

There is so far not much experience with the method and the advantages of simplicity might be offset by the out-of-control risks. Wahrendorf and Brown (1980) report an investigation on comparing the actions on two drugs and Rubin (1981) proposes to introduce a bayesian viewpoint into the bootstrap method. In technical reports, Efron (1980, 1981a) compares the bootstrap method with quite a few parent methods; his findings favor the bootstrap method but this may be due to the selection of estimators he investigated.

PART B

In this second part we meet with a few classical problems, which are well-known in the theory of statistical estimation. However the viewpoint is rather unusual and this leads to disregard certain aspects which otherwise would be of great concern.

For example, consider the estimation of a location parameter ϑ for the variate x distributed according to the law $f(x-\vartheta)$. We will search for the "best" estimator without feeling concerned by its representativity or its real meaning. We may end up with a median estimate or an arithmetic mean; both are invariant with respect to the real location under translation of the distribution; both may frequently be used to estimate the location, however they may differ.

Chapter 6
Type M estimators

6.1. Definition

The type M estimators, also called M estimators, are generalizations of the usual maximum likelihood estimates. ϑ is classically the parameter value maximizing the likelihood function, i. e. we have in obvious notation

$$L = \prod f(x_i \mid \vartheta) = \max \text{ for } \vartheta$$

or equivalently

$$-\ln L = -\sum \ln f(x_i \mid \vartheta) = \min \text{ for } \vartheta .$$

The estimators of type M are solutions of the more general structure

$$M = \sum \rho(x_i, \vartheta) = \min \text{ for } \vartheta , \tag{6.1}$$

where the function $\rho(.)$ may be rather arbitrary. Before proceeding, we note that the above structure is rarely appropriate to process correlated observations.

The estimators of type M have been analyzed in quite many respects for location problems, since the initial contribution of Huber (1964) was published; however, the many other estimation problems have received scant attention. A natural extension of the structure (6.1) is obtained by considering systems of functions to be minimized, each of these functions relating to a given estimator and the estimators having some inter-dependence; this is treated in Chapter 9 on MM estimators. For the time being we restrict ourselves to the problem (6.1) or rather to its weighted form,

$$M = \sum w_i \rho(x_i, \vartheta) = \min \text{ for } \vartheta , \tag{6.2}$$

where

M is a scalar,

w_i are non-negative weights,

$x_i \in \Omega \subset R^p$,

$\vartheta \in R^q$,

$\rho(.,.)$ is a scalar function.

In order to place ourselves in the setting of Chapter 2, we also write (6.2) in a form which can be regarded in two ways, on the one hand as the definition of the parameter and, on the other hand, as the definition of the parameter estimate, namely

$$M = \int_\Omega \rho(x,\vartheta)\, f(x)\, dx = \min \text{ for } \vartheta \ . \tag{6.3}$$

When $f(x)$ is the true (unknown) probability distribution, (6.3) constitutes a definition of ϑ; when $f(x)$ is the empirical distribution (2.40), i. e.

$$f(x) = (\textstyle\sum w_i)^{-1} \sum w_i\, \delta(x-x_i) \ , \tag{6.4}$$

(6.3) becomes an estimation rule defining the estimator $\hat{\vartheta}$ of parameter ϑ.

Throughout we assume suitable differentiability with respect to the ϑ-components. More precisely, we assume

- Independence of Ω on ϑ (conditions on $f(x)$ at the frontier of the sample space Ω generally meet the needs)

- Existence and (piecewise) continuity of the derivatives

$\dot{\rho}(x,\vartheta) = (\partial/\partial\vartheta)\, \rho(x,\vartheta)$ - which is a vector

$\ddot{\rho}(x,\vartheta) = (\partial/\partial\vartheta)\, \dot{\rho}(x,\vartheta)$ - which is a matrix.

Then we may also replace the minimization (6.3) by the equivalent form

$$\int_\Omega \dot{\rho}(x,\vartheta)\, f(x)\, dx = 0 \tag{6.5}$$

with

$$A = \int_\Omega \ddot{\rho}(x,\vartheta)\, f(x)\, dx , \ positive\ definite \ . \tag{6.6}$$

It will be noted that (6.6) is a little too constraining when requiring A to be positive definite at the solution ϑ of the implicit equation (6.5). In some circumstances, a compact set of solutions ϑ may result from (6.3) and then $A = 0$; however an evaluation of A on the frontier of the open set complementary to the solution set should yield a positive definite matrix. These are technical details which can be of importance when working with discontinuous $\dot{\rho}(x,\vartheta)$; in what follows we will tolerate discontinuities with respect to ϑ as long as they take place in a zero-measure set of R^g, the space of the ϑ-variable. The above makes it clear what we mean by requiring "piecewise" continuity of the first two ρ-derivatives; further details are given in Huber (1977c, pp. 13-16).

6.2. Influence function and variance

For further use, it is convenient to expand (6.5) in series in the vicinity of the solution ϑ. Let ϑ^* be a vector-value close to ϑ, then

$$\int \dot{\rho}(x,\vartheta)\, f(x)\, dx = \int \dot{\rho}(x,\vartheta^*)\, f(x)\, dx + A(\vartheta-\vartheta^*) = 0 \qquad (6.7)$$

where the matrix A is to be estimated by (6.6) at a point intermediate between ϑ and ϑ^* (see 2.30). However, in view of the piecewise continuity of A, the exact evaluation point should not matter too much. In practice, an estimator $\hat{\vartheta}$ will be implicitly defined by

$$(\textstyle\sum w_i)^{-1} \sum w_i\, \dot{\rho}(x_i,\vartheta^*) + A(\vartheta-\vartheta^*) = 0 \qquad (6.8)$$

where A stands for

$$A = A(\vartheta^*) = (\textstyle\sum w_i)^{-1} \sum w_i\, \ddot{\rho}(x_i,\vartheta^*)\; ; \qquad (6.9)$$

being a generalization of the usual information matrix. These two expressions result from the introduction of (6.4) into (6.6-7). They immediately lead to an estimation scheme which might be appropriate for ϑ^* close to the solution ϑ. It consists in improving a previous estimate of ϑ, here denoted by ϑ^*, by iterating at will on

$$\vartheta = \vartheta^* - [\textstyle\sum w_i\, \ddot{\rho}(x_i,\vartheta^*)]^{-1} \sum w_i\, \dot{\rho}(x_i,\vartheta^*)\; . \qquad (6.10)$$

Let us now embark on estimating the influence function of Hampel (1974) at § 3.3. This can be obtained by direct application of the Gateaux's derivative definition onto (6.5). Let ϑ be the solution with respect to $f(x)$ as well as ϑ^* be the solution with respect to the perturbed problem where the probability density is

$$f^*(x) = (1-t) f(x) + t\delta(x-x_0) .$$ (6.11)

Then, according to (2.39), the influence function in x_0 is

$$\Psi(x_0) = \lim_{t \to 0} (\vartheta^* - \vartheta)/t .$$ (6.12)

It is easily derived as indicated by the following relations. By definition

$$\int \dot\rho(x,\vartheta^*) f^*(x) \, dx = 0$$

and, confining ourselves to the first order term in the perturbation,

$$\int \dot\rho(x, \vartheta^*) f^*(x) \, dx$$

$$= \int [\dot\rho(x, \vartheta) + \ddot\rho(x, \vartheta)(\vartheta^* - \vartheta)] [(1-t) f(x) + t\delta(x-x_0)] \, dx$$

$$= \int \ddot\rho(x, \vartheta) (\vartheta^* - \vartheta) f(x) \, dx + t \dot\rho(x_0, \vartheta)$$

$$= A(\vartheta^* - \vartheta) + t \dot\rho(x_0, \vartheta) = 0 .$$

Eventually (6.12) yields the influence function

$$\Psi(x_0) = - A^{-1} \dot\rho(x_0, \vartheta)$$ (6.13)

where A is defined by (6.6) or (6.9).

Deriving the variance-covariance matrix (or the third centered moment) of an M estimator is now a trivial matter. By the use of (6.13), (4.31) and (4.35), we obtain

$$cov(\vartheta) = [(\textstyle\sum w_i)/\overline{w} - 1]^{-1} A^{-1} B A^{-1}$$ (6.14)

where

$$B = \int [\dot{\rho}(x,\vartheta)] \, [\dot{\rho}(x,\vartheta)]' \, f(x) \, dx$$

$$= (\textstyle\sum w_i)^{-1} \sum w_i \, [\dot{\rho}(x_i,\vartheta)] \, [\dot{\rho}(x_i,\vartheta)]' \tag{6.15}$$

and, according to (4.37),

$$\overline{w} = \sum w_i^2 / \sum w_i \ . \tag{6.16}$$

In order to grasp the structure and implications of (6.14), let us see what comes out in the ordinary set up of maximum likelihood estimation. Then, except for an arbitrary non-zero multiplicative constant, we have in view of (6.1)

$$\rho(x_i,\vartheta) = -\ln f(x_i \mid \vartheta) \ .$$

Substituting this in (6.9), we obtain in the equal weight situation

$$A = \frac{1}{n} \sum_i \frac{\partial^2}{\partial \vartheta^2} \, [-\ln f(x_i \mid \vartheta)]$$

which closely approximates the total Fisher information matrix as given by the expected value

$$F_\vartheta = Expec \, \{ - \frac{\partial^2}{\partial \vartheta^2} \ln f(x_i \mid \vartheta) \} \ .$$

When this formula is used for the maximum likelihood solution ϑ and for the sample at hand, A can be considered as the "observed" information matrix. Similarly (6.15) yields

$$B = \frac{1}{n} \sum_i \, [\frac{\partial}{\partial \vartheta} \ln f \, (x_i,\vartheta)] \, [\frac{\partial}{\partial \vartheta} \ln f \, (x_i,\vartheta)]'$$

which happens to be just another form of the information matrix. The merits and drawbacks of "observed" information (matrices) are nicely presented by Efron and Hinkley (1978) and Efron (1978), although they have been confronted by stiff opposers in the discussion of their paper. Hence, in the maximum likelihood set-up, we obtain from (6.14)

$$cov(\vartheta) \approx [n \, F(\vartheta)]^{-1}$$

and this opens directly the door to the various concepts surrounding the Cramer-Rao type inequalities. We will not go into them, but we should like to recommend two relevant papers by Kozek (1977) and Khatri (1980) who significantly enlarge the usual scope.

To conclude this section we should like to add two further remarks: on the one hand, the equivalence of A and B is a by-product of the maximum likelihood approach and is exceptional; on the other hand, it makes sense to work out a variance-covariance matrix only if the estimator is more or less normally distributed (what would this mean if not?).

6.3. Robust M estimators

Apart from having a bounded influence function (see § 3.3), it is of course quite natural to require robust estimators to be unique. This implies that the function M of variable ϑ to be minimized in (6.2) should have a unique minimum. This is a constraint on a weighted sum of ρ-functions which is necessarily satisfied when

$$\rho(x, \vartheta) \text{ is convex in variable } \vartheta . \tag{6.17}$$

It will be noted that (6.17) is too constraining, but it seems that it is hardly possible to avoid the convexity criterion. For instance, it is evident that the sum of functions with a unique minimum may have several minima. This is the case with maxima when considering mixture distributions; the sum of unimodal probability distribution is very often multimodal. Hence to guarantee a unique solution we impose convexity or, equivalently, we impose

$$\ddot{\rho}(x,\vartheta) , \text{ non-negative definite} \tag{6.18}$$

whatever are x and ϑ; this is consistent with (6.6). The following further requirement may be regarded as minor in spite of its very great practical importance. We not only want a unique solution, but we would also appreciate to be in a position to find it without having to search through the complete g-dimensional space of ϑ (well designed

95

numerical procedures can cope with most situations). This leads to the practical requirement that $M(\vartheta)$ should have a gradient whenever A by (6.9) is singular. With the same argument as used in the above discussion, this implies the less important requirement

$$\dot{\rho}\,(x,\vartheta) \neq 0\,,\text{ if }\ddot{\rho}(x,\vartheta)\ \textit{is singular}\ . \tag{6.19}$$

Before leaving this convexity aspects, note that they have placed us in a context where the theory is well advanced. For instance it seems that much more can be derived from Jensen's inequality than is usually done in the multivariate space (e. g. see Schaeffer, 1976, Das Gupta, 1980 and Eaton, 1982).

Let us now proceed to discuss robustness in the sense of limited sensitivity to each individual observation. As pointed out in § 3.3, the influence function must be suitably bounded but, in view of (6.13), is not at all clear how it should be bounded. Many types of norms $||\Psi(x)||$ could be of interest but it can be conjectured that the most appropriate one is the ordinary euclidean norm. We will return to it presently.

Assume that we have obtained an influence function which is bounded whatever x is. Then, according to (2.41), an estimator $\hat{\vartheta}$ of parameter ϑ is the sum of ϑ and many bounded terms. Since we satisfy the conditions of the central limit theorem, we may infer that the tendency towards a multivariate normal law is fairly rapid. This has been investigated in more general terms than required by the robustness framework by quite a few authors; let us mention among the recent papers those of Hartigan (1975), Landers and Rogge (1976), Major (1978), Hall (1980) and Sweeting (1980).

Therefore we may expect the estimator $\hat{\vartheta}$ to have an approximately normal distribution and this recommends the variance-covariance matrix as a way of assessing the estimator scatter. We now show that minimizing $Cov(\vartheta)$ as we always try to accomplish is consistent with the use of the euclidean norm for the influence function. Inasmuch as minimizing the matrix $Cov(\vartheta)$ is equivalent to minimizing its trace, we have for the asymptotic variance v by (4.31) and (4.35)

$$V = \left[\sum w_i / \overline{w} - 1\right] Cov\,(\vartheta)$$

$$= Ave\,\{[\Psi(x_i)]\,[\Psi(x_i)]\,'\}$$

$$= Expec\,\{[\Psi(x)]\,[\Psi(x)]\,'\}\;. \qquad (6.20)$$

$$tr\,\{V\} = Expec\,\{tr\{[\Psi(x)]\,[\Psi(x)]\,'\}\}$$

$$= Expec\,\{[\Psi(x)]\,'\,[\Psi(x)]\,\}$$

$$= \int\,||\Psi(x)||^2\,f\,(x)\,dx\;.$$

Applying the euclidean norm definition, it is simple although rather lengthy to obtain the following results:

a. The asymptotic variance v by (6.20) is minimum for $\rho(x,\vartheta)$ in (6.1) proportional to $[\ln f\,(x\,|\vartheta)]$. This result confirms the optimality of the maximum likelihood estimators; we have not yet learned so much.

$$(6.21)$$

b. A minimum variance robust estimator such that

$$||\Psi(x)||\,\leq c_1$$

can be obtained by the following iterative algorithm

- Define a function $\rho_1(x,\vartheta)$ proportional to the log-likelihood, e. g.

$$\rho_1(x,\vartheta)\,=\,-\ln f\,(x\,|\vartheta)\;.$$

This function will be the initial function $\rho(x,\vartheta)$ to be considered in this algorithm. The function $\rho(x,\vartheta)$ obtained after convergence will be the solution to be inserted in (6.2).

- Given the (approximate) function $\rho(x,y)$, work out the information matrix A by its definition (6.6)

$$A\,=\,\int \ddot{\rho}(x,\vartheta)\,f\,(x)\,dx\;.$$

- In the g-dimensional space of ϑ (and for any given x), define the g-dimensional subset S_- such that the bounding condition is met. The complementary subset is S_+; i. e.

$$S_- = \{\vartheta : ||\,\Psi(x)||\, \leq c_1\}$$

and

$$S_+ = \{\vartheta : ||\,\Psi(x)||\, > c_1\}$$

with

$$\Psi(x) = -A^{-1}\dot{\rho}\,(x,\vartheta)\ .$$

Observe that an empty S_- indicates that there is no solution corresponding to so small a constant c_1.

- Construct the final function $\rho(x,\vartheta)$ in such a way that it meets the conditions

$$\rho(x,\vartheta) = \rho_1(x,\vartheta),\ \text{if}\ \vartheta \in S_-$$

$$\dot{\rho}(x,\vartheta)\ \text{continuous},\ \vartheta \in S_- \cup S_+$$

$$||\Psi(x)|| = c_1,\ \vartheta \in S_+$$

The last two conditions define $\rho(x,\vartheta)$ as being a certain continuation in S_+ of the values taken in S_-. How these continuations behave needs further investigation.

- Work out a new matrix A by

$$A = \alpha A + (1-\alpha) \int \dot{\rho}\,(x,\vartheta)\,f(x)\,dx$$

where the right-hand side A is the previous matrix A and the parameter α controls the convergence $(0 \leq \alpha < 1)$.

- Repeat the algorithm from the last step but two until it has converged. The organization of the algorithm may require modifications to speed up the convergence.

$$(6.22)$$

Both results (6.21) and (6.22) are multivariate extensions of the derivation presented in Rey (1978, pages 49-57) and it moderately supports the conjecture in Huber (1972, §12.3) concerning robust maximum likelihood estimators. It will be observed that both results imply knowledge of $f(x)$ and that we have implicitly assumed that difficulties with respect to multiple solutions did not occur; any constraints (6.17-19) have been ignored. Rey (1978, pages 57-59) proposes an application of (6.22). It must be observed that the algorithm (6.22) is a combination of standard statistical theory (the likelihood function) and mathematical features imposed for extraneous reasons; the latter could be regarded as slightly bayesian in being a way of implementing prior ideas on how the solution should be sensitive to the observation (see Dempster, 1968). Before leaving the question of how to design an estimation procedure with an influence function that is bounded everywhere, let us mention that Hampel (1978) presents a few conjectures which could provide a better control in the parameter space. By contrast note that Krasker and Welsch (1982) concentrate their approach on a weighting of the sample space. The various views are confronted in the discussion paper of Huber (1983). Among the discussants, Mallows states that "many of the estimators ... violate at least one of Hampel's intuitive desiderata for robust estimates". Thus far we are unclear about some of the issues.

A robust estimator ϑ in a regression model $y_i = x'_i\vartheta + \varepsilon_i$ is defined through the mathematical rule of minimization rather than through a statistical characterization of the possibly-random variable ε. Inasmuch as $\rho(x_i, \vartheta)$ is an increasing function of $|\varepsilon_i|$, the minimization process can also be looked upon as a way of approximation y_i by $x'_i\vartheta$, in spite of possible inadequacy of the linear model.

In location, the emphasis on the mathematical rule rather than on the statistical characterization has been skipped by Huber (1964), and partly by Jaeckel (1971), by considering symmetric distribution $f(x)$ where there is a "natural" definition of ϑ, the center of symmetry. Then limiting $\rho(.)$ to be symmetric resolves most statistical problems.

This emphasis underlies the discussion of what is robustness (or what it should be) by Huber (1972, 1973, 1977a). The minimization viewpoint is more critically analyzed in more recent investigations into regression problems; Jaeckel (1972b) and Collins (1976) state their concerns. But rather generally, authors ignore the possibility of a statistical characterization when they study finite samples and only take in consideration asymptotic properties, e. g. Maronna (1976). In most cases the finite sample properties are conjectured from Monte Carlo simulations. A nice case in the literature is Hampel's papers (1973a, 1975) advocating the use of appropriate mathematical rules, which have been directly opposed by Dempster (1975, 1977) who only works on the statistical characterization in a bayesian framework. Obviously if one admits to select a data-dependent prior, very good and robust estimators can be derived. The dependence on data is often introduced sequentially (the prior is modified until "satisfactory" results are obtained) and frequently consists in arbitrary trimming or winsorization, see Yale and Forsythe (1976). This may require a specific identification of outlying observations and many methods have been investigated in this regard since Anscombe (1960); we will meet this topic in Chapter 11.

However we think necessary to endure values, as does Youden (1972), in keeping a critical eye on what is produced, as Mead and Pike (1975) recommend.

After having traveled a long way between estimation regarded as a statistical procedure or as a type of approximation, let us resume our development of these particular type M estimators.

6.4. Robustness, quasi-robustness and non-robustness

Thus far we have been rather rigorous in the derivations relating to robustness. One of the outcomes has been the preceding section where we have dealt with estimators which provide unique solutions, are robust (in the sense of bounded influence function) and are minimum variance. Two aspects have been entirely ignored: First, do we need them? Second, are they tractable? The answer to the first question is a definite "No"; the second question cannot be answered in fully general terms for, depending upon the specific situation, the robust estimators can be worked out with more or less effort.

Thus the most important issue is whether we really need robust estimators. Let us place ourselves in the very frequent situation in which a statistician has been called for a given job. The details of the analysis he makes has no meaning whatsoever for the principal who usually is unable to understand any advanced statistical method. He expects good work as well as results he can understand and trust. Good work and reliable results imply in most cases the use of techniques which are more robust than are the standard ones. Just to give a specific example, all techniques of linear statistical inference assume that the errors are normally distributed (when not even i. i. d.); whether the data set fits in such a constraining model is very difficult to assess and, when it does not, what to do is hardly known. Hence, although without any justification, the statistician analyzes the data set with a model of normally distributed errors; he may do so if he checks whether the errors do affect the results or not. This check on the errors cannot be performed directly, in fact he may at most have the disposal of a set of residuals to which he must apply tests; however all tests on residuals have low powers due to the fact that large errors have been spread into (at least a few) residuals. By contrast the robust methods yield solutions which are not sensitive to error distributions and, with them, similar solutions should be obtained. Any marked discrepancy between the ordinary solutions and robust solutions is indicative of trouble: the ordinary solutions were not reliable.

We define **quasi-robust solutions** *as being much more robust than ordinary solutions without being strictly robust. The corresponding quasi-robust methods are suitable for comparison with the ordinary methods in order to ascertain method reliability. They are improvements on the ordinary methods without being as demanding in computing facilities as are* *robust methods.*

In the last part of this section, we consider some of the main proposals which have been advanced for the $\rho(x,\vartheta)$-function in (6.2). As all
these proposals have been designed in the set-up of one-dimensional
location problems, we will here restrict them to cover the same scope;
the generalization to several dimensions is covered by Collins (1982).
Much more general use will be presented in other chapters. It will be
noted that we assume to have prior knowledge of any further parameter, such as a scatter measure for instance.

6.4.1. Statement of the location problem

Based on a sample $\{x_1,...,x_n\}$ with (possibly equal) weights $\{w_1,...,w_n\}$ and
drawn from the probability distribution $f(x-\vartheta)$, we estimate ϑ by (6.2)
where

$$\rho(x,\vartheta) = \rho(x-\vartheta) . \tag{6.23}$$

It is assumed that $f(x-\vartheta)$ has a unit scatter measure (scale = 1). All
functions (6.23) depend upon a parameter which can be adjusted
according to specific wishes; this permits the function to be tuned with
what is referred to as the tuning parameter as do Coleman et al (1980).
Varying the tuning parameter yields families of functions and, hence,
families of estimators; these estimators do differ in efficiency and this
feature will be used in order to compare estimators which belong to
different families or, equivalently, which are produced by different ρ-
functions. In Table 6 we have listed the values of the tuning parameters
such that relative asymptotic efficiencies 0.95, 0.90, 0.85 and 0.80 are
obtained, the relative asymptotic efficiencies are evaluated under the
assumption of normally distributed errors. Conclusions will be found in
§ 6.4.10.

6.4.2. Least powers

This is the only class of ρ-functions which do not depend upon any prior knowledge of a scatter measure. It may be shown that this feature makes them non-robust (see Rey 1978, p. 43). In this class, we have

$$\rho(x,\vartheta) = |x-\vartheta|^{\nu} , \nu > 1 . \tag{6.24}$$

The constraint $\nu > 1$ results from (6.17).

In this family of $\rho(x-\vartheta)$ there are found several well known schemes. The most famous method is certainly the least squares minimization with $\nu = 2$, but the Chebyshev's criterion and the absolute value minimization, respectively ν tending to infinity and to 1, have also received much attention.

In what follows, large values of the parameter ν will not be considered because they are of no concern for the common statistical applications. In fact the smaller ν, the smaller is the incidence of large residuals in the ϑ-estimate; it appears that ν must be fairly moderate to provide a relatively robust estimator or, in other words, to provide an estimator scarcely perturbed by outlying observations. The selection of an optimal ν is investigated by Ronner (1977, Ch. 4).

The value $\nu = 2$, corresponding with the least squares or the arithmetic mean, is still too large, as we can infer from the opinion of the Princeton group based on Monte Carlo runs. In Andrews et al. (1972, p. 239), they answer the question

Which was the worst estimator in the study? If there is any clear candidate for such an overall statement, it is the arithmetic mean, long celebrated because of its many "optimality properties" and its revered use in applications. There is no contradiction: the optimality questions of mathematical theory are important for that theory and very useful, as orientation points, for applicable procedures. If taken as anything more than that, they are completely irrelevant and misleading for the broad requirements of practice. Good applied statisticians will either look at the data and set aside (reject) any clear outliers before using the "mean" (which, as the study shows, will prevent the worst), or they will switch to taking the median if the distribution looks heavy-tailed.

The difficulties have been surveyed by Huber (1972) and it may be concluded that only a limited importance must be attached to the most extreme data. For ν around 1.2, a good estimate may be expected, but this is a matter of opinion.

Technically the only parameter value which bounds the large residual incidence is $\nu = 1$. But then strict convexity is lost and an indeterminate solution may result. The corresponding problems are well known and are usually solved by linear programming techniques. However the indeterminacy may also be resolved by considering $\nu = 1$ as the lower limit of $\nu > 1$. If ϑ does not vary much with ν, this procedure appears reasonable; it has been recommended by Jackson (1921) who investigated the one-dimensional median. - Our own experience entirely supports his view. - An interesting aspect is that this provides a natural extension of the one-dimensional median to the multivariate domain. The dependency of ϑ on ν has been given great attention in view of the current computational facilities. Nowadays, we can compute, but is it meaningful in theory? The answer is positive but reluctant for most authors, a reluctancy related to the relevance of the requisite theoretical assumptions. Fletcher et al. (1974) approach the problem from a computation oriented viewpoint, whereas Cargo and Shisha (1975), Hwang (1975) as well as Lewis and Shisha (1975) analyze topological aspects. The theoretical considerations have a serious impact on practice because a great number of tricks are used in the algorithms to force their eventual convergences to possibly artificial solutions. Before leaving these considerations, note that instead of minimizing the sum of powers of residuals an immediate generalization permits us to minimize vector norms and thus to extend ϑ from a vector to a matrix structure. These norm minimizations are investigated by Boyd (1974) and Rey (1975a); the latter paper is partially concerned with multidimensional location problems.

As said, many difficulties are encountered in the computations when the parameter ν is in the range of interest $1 < \nu < 2$, because zero residuals are troublesome.

Relevant Constants for some Rho-functions

E	Tuning constant	$\psi(1)$	$\psi(2)$	$\psi(3)$	$\psi(3)/\psi(1)$	β (§ 6.4.5)	Int (9.20)
Least powers (6.24)							
0.95	1.5788	1.1406	1.7035	2.1541	1.8886		0.8768
0.90	1.4244	1.1824	1.5867	1.8847	1.5940		0.8458
0.85	1.3123	1.2081	1.5000	1.7025	1.4093		0.8278
0.80	1.2216	1.2255	1.4289	1.5633	1.2756		0.8160
Function of Huber (6.25)							
0.95	1.3450	1.2175	1.6375	1.6375	1.3450	0.4576	0.9326
0.90	0.9818	1.4571	1.4571	1.4571	1.0000		0.8431
0.85	0.7317	1.3660	1.3660	1.3660	1.0000		0.7338
0.80	0.5294	1.3121	1.3121	1.3121	1.0000		0.6035
Modified Huber function (6.26)							
0.95	1.2107	1.2229	1.6634	1.6634	1.3602	0.4816	0.8511
0.90	0.8590	1.3456	1.4651	1.4651	1.0888		0.7512
0.85	0.6319	1.3691	1.3691	1.3691	1.0000		0.6436
0.80	0.4538	1.3133	1.3133	1.3133	1.0000		0.5233
Fair (6.27)							
0.95	1.3998	1.1866	1.6751	1.9416	1.6363	1.0176	0.6015
0.90	0.6351	1.2383	1.5368	1.6710	1.3494		0.4239
0.85	0.3333	1.2608	1.4409	1.5129	1.2000		0.2915
0.80	0.1760	1.2677	1.3702	1.4082	1.1108		0.1874
Function of Cauchy (6.28)							
0.95	2.3849	1.2195	1.6838	1.6659	1.3660	0.3629	0.8228
0.90	1.7249	1.3176	1.5018	1.3121	0.9959		0.7295
0.85	1.3737	1.3847	1.3582	1.1016	0.7956		0.6516
0.80	1.1385	1.4298	1.2398	0.9566	0.6690		0.5824
Function of Welsch (6.29)							
0.95	2.9846	1.2111	1.7297	1.4801	1.2221	0.2378	0.8580
0.90	2.3831	1.3185	1.5549	0.9670	0.7334		0.7952
0.85	2.0595	1.4102	1.3903	0.6416	0.4550		0.7450
0.80	1.8383	1.4939	1.2298	0.4201	0.2812		0.7009
Bisquare (6.30)							
0.95	4.6851	1.2021	1.7650	1.3780	1.1463	0.1969	0.8737
0.90	3.8827	1.3051	1.6161	0.7294	0.5589		0.8230
0.85	3.4437	1.3970	1.4634	0.2905	0.2080		0.7825
0.80	3.1369	1.4856	1.2968	0.0403	0.0271		0.7465
Function of Andrews (6.31)							
0.95	1.3387	1.2022	1.7642	1.3868	1.1535	0.1937	0.8726
0.90	1.1117	1.3037	1.6218	0.7138	0.5475		0.8225
0.85	0.9884	1.3938	1.4784	0.1744	0.1251		0.7829
0.80	0.9022	1.4810	1.3213	0.0000	0.0000		0.7479

Table 6

Several well known methods are available to minimize functions. Let us first discard all the methods relying on some separation of function components (e. g. orthogonal decomposition); due to the fairly involved non-linearities, it does not appear possible for us to proceed in this way for the parameter ν in the range of interest. The search techniques of minimization must also be rejected because they have a very poor accuracy and may give numerical difficulties for a "flat" function minimum, a situation which quite often occurs. Thus we are left with relaxation and gradient methods. The former, as used by Gentleman (1965), are equivalent to first order gradient methods in this context and will be presented in §9.8.1 for MM estimators.

Unfortunately, the second order gradient algorithms cannot be considered due to the problems arising in the determination of the second order derivatives. They involve the absolute residuals, $|x_i - \vartheta|$, to a negative power and they are responsible for hazards in the algorithm convergence whenever the solution ϑ correspond to small residuals. To avoid this difficulty, Forsythe (1972) recommends the implementation of the Davidson method (Fletcher and Powell, 1963), whereas Ekblom (1973, 1974) proposes the elimination of the poles by a quadratically perturbed method. It is noteworthy that the Ekblom method corresponds to a somewhat restricted case of the generalized problem of Rey (1975a). Indeed, it consists in the addition of a fictitious second dimension to the scalar data, u_i. We have preferred to apply the first order gradient method with a specially chosen step size, that is, chosen in order to "avoid" the poles of the second derivatives. A discussion of the convergence is given by Wolfe (1979) and the idea of adding a fictitious further dimension is also advanced by Weber and Werner (1981). However, whatever the algorithm, the computational burden remains substantial.

With regard to computational experience, Merle and Spath (1974) present a clear discussion of several methods to be found in the literature. Comparison of the least powers approach with other techniques has been proposed by Rey (1977).

According to (6.21), the least power function (6.24) is optimal for the probability distribution of density

$$f(x) = \nu\,[2\Gamma(1/\nu)]^{-1}\,\exp(-|x|^{\nu})$$

which for increasing parameter ν goes from the Laplace distribution to the normal distribution and eventually to very light tail laws. The asymptotic efficiency of the least power function applied to the normal law is according to (6.20) given by

$$2(\pi)^{-1/2}\,\{\Gamma[\,(\nu+1)/2\,]\}^2\,/\,\Gamma(\nu-1/2)$$

This expression has the value 0.95 for the tuning constant $\nu = 1.5788$. The relatively low value $\nu = 1.2$ we have recommended yields the asymptotic efficiency 0.79 whereas the value for median estimation ($\nu = 1$) would be 0.64. This last estimator has been examined in very many situations; a good analysis in a realistic set-up has been reported by Hill and Holland (1977). However, because of the moderate performance of the least absolute residual approach as well as its lack of unicity, we hesitate to apply such a method.

6.4.3. Huber's function

Assume for a moment we intend to design the most simple function we can imagine which is consistent with the robustness conditions. It happens that this ρ-function is precisely the one Huber (1964) has claimed to be optimal; it is the minimax solution for the location problem with known scale parameter in the case of contaminated normal distributions.

When the distribution of the sample is unknown, or vaguely known, the most plausible assumptions consist in supposing symmetry

$$f(x-\vartheta) \approx f(\vartheta-x)$$

and validity of a Taylor expansion in the vicinity of its mode, namely

$$f(x) \approx f(\vartheta) - a(x-\vartheta)^2 \ .$$

Recalling (6.21), this leads to a function in the vicinity of the mode ϑ

$$\rho(x-\vartheta) \approx -b \ln [f(\vartheta) - a(x-\vartheta)^2]$$

$$\approx -b \ln f(\vartheta) + [a b / f(\vartheta)] (x-\vartheta)^2$$

where a and b are positive constants. Hence a parabola

$$\rho(x-\vartheta) = (x-\vartheta)^2$$

is the optimal selection in the vicinity of the mode. However we have just seen that the least squares approach is not robust. This is because the influence function is not bounded; a parabolic function is increasing too rapidly. We proceed as in the algorithm (6.22); for some given coordinate

$$|x-\vartheta| = k$$

we limit the influence function at the level it has acquired. This leads to the so-called Huber's proposal 2:

$$\rho(x-\vartheta) = \rho(u) = \begin{cases} u^2 , & \text{if } |u| \leq k \\ k(2|u|-k) , & \text{if } |u| \geq k \ ; \end{cases} \tag{6.25}$$

There are no more words to add to this ρ-function for all its aspects have been discussed by many authors. It is so satisfactory (in spite of the various incredible uses made of it) that it has been recommended for almost all situations; very rarely it has been found to be inferior to some other ρ-function but that was in the case of unrealistic problems. In this context we only refer to the basic work of Huber (1964, 1972), because (6.25) appears under one form or another in all the literature on robustness.

The 0.95 asymptotic efficiency on the standard normal distribution is obtained with the tuning constant k = 1.3450.

6.4.4. Modification to Huber's proposal

As we have said above the function (6.25) gives remarkable results; however from time to time difficulties are encountered in working them out. This may occur in spite of great care in the design (or selection) of the algorithm. Our personal experience has shown that the troubles can always be attributed to a lack of stability in the gradient values of the ρ-function - we use the term of gradient rather than mentioning the first derivative to anticipate more involved estimation problems - Indeed, (6.25) has a discontinuous second derivative,

$$\ddot{\rho}(x_i - \vartheta) = \begin{cases} 2, & \text{if } |x_i - \vartheta| < k \\ 0, & \text{if } |x_i - \vartheta| > k \end{cases}$$

hence A in (6.9) exhibits a discontinuous behavior while ϑ varies during the algorithmic iterations. The modification we now propose is such that the influence of an observation on the second derivative changes gradually as ϑ varies

$$\rho(x - \vartheta) = \rho(u) = \begin{cases} 2c^2[1 - \cos(u/c)], & \text{if } |u|/c \leq \pi/2 \\ 2c|u| + c^2(2-\pi), & \text{if } |u|/c \geq \pi/2 \end{cases} \tag{6.26}$$

can also be regarded as similar to Andrew's proposal (6.31). The essential difference is that the limit is $\pi/2$ rather than π, which guarantees a unique solution of (6.17).

The 0.95 asymptotic efficiency on the standard normal distribution is obtained with the tuning constant c = 1.2107. It is interesting to note that observations satisfying

$$|x_i - \vartheta| < c\pi/2 = 1.9018$$

occur in the strictly convex part of (6.26), whereas the corresponding interval is 30% smaller for Huber's proposal (6.25) which satisfies

$$|x_i - \vartheta| < k = 1.3450.$$

Inspection of the other features reported in Table 6 reveals that, on all other grounds, there is no large difference between (6.25)- and (6.26)-performances.

6.4.5. Function "Fair"

We have not found who has introduced this very remarkable ρ-function. Indeed it exhibits a very nice property which, as far as we know, has never been noted. Furthermore, it has a very simple structure which is so simple that it could be considered as a reason to link the function Fair" to the fundamental particles "Charm" and "Beauty"; what lovely names! This function is among the possibilities offered by the Rosepack package of Klema (1978) and Coleman et al (1980).

Let us abandon for the time being what we have just discussed and consider a more general set-up. Rather than dealing with the one-dimensional location problem with known unit scatter, we may have to analyze data sets without any prior knowledge of the error scale. Then the scale must also be estimated, let it be denoted by s. Then

$$\rho(x_i, \vartheta) = \rho\left[(x_i - \vartheta)/s\right]$$

will be treated as being relative to variates with unit scatter. The trouble is that the scale s may be very difficult to estimate reliably. Therefore we would prefer not to depend upon such an estimate. This situation occurs in linear regression, and even to a greater extent in problems in which the regression is deemed linear rather than definitely linear. In standard notation, the estimator is denoted by β and appears in a ρ-function of the form

$$\rho\left[(y_i - x'_i \beta)/s\right]$$

where s is some "average" scale; when the homoscedasticity is not guaranteed, the scale parameter s is overestimated for some observations and underestimated for some others. Thus, seeing that homoscedasticity should be ruled out whenever robust methods are invoked, it is preferable to have a low sensitivity to the scale parameter value or, in other words, to the value of the tuning constant. This is the most interesting feature of the function "Fair", namely

$$\rho(x-\vartheta) = \rho(u) = 2c^2 \left[|u|/c - \ln(1 + |u|/c) \right] . \tag{6.27}$$

This ρ-function yields robustness as well as unicity. The 0.95 asymptotic efficiency on the standard normal distribution is obtained with the tuning constant $c = 1.3998$.

In order to assess the sensitivity of the relative efficiency to the tuning constant value, we have to design a sensitivity index; this will now be done. The tuning constant can be regarded as a function of the required efficiency, thus

$$c = c \, (efficiency)$$

and, in virtue of the manner we have parametrized functions (6.25-31), we always obtain the following correspondences

$$c = 0 ==> E = 0.63662 ,$$

$$c = \infty ==> E = 1 ,$$

where E stands for the relative asymptotic efficiency. This leads to a crude approximation of the tuning constant

$$c = c(E) \approx \alpha \left[(E - 0.63662)/(1-E) \right]^\beta .$$

The constants α and β can be immediately derived from the values taken for two different efficiencies. The values of the parameter β are given in Table 6 as they have been derived from the two sets (c(0.95), E = 0.95) and (c(0.8), E = 0.8). In the present situation, we obtain $\beta = 1.02$, a value at least two times higher than any other index.

6.4.6. Cauchy's function

This is the first ρ-function we meet with which does not guarantee a unique solution. With a descending first derivative, $\dot{\rho}(x-\vartheta)$, such a function has a tendency to yield erroneous solutions in a way which cannot be observed. There is no way to find out whether the proposed solution

111

is valid except by running a more reliable ρ-function. This shortcoming has been reported by Rey (1977).

The following ρ-function has been named according to its optimality for the Cauchy distribution:

$$\rho(x-\vartheta) = \rho(u) = c^2 \ln\left[1 + (u/c)^2\right].$$

(6.28)

The 0.95 asymptotic efficiency on the standard normal distribution is obtained with the tuning constant $c = 2.3849$. With an index $\beta = 0.36$, it happens to be the best among all non-admissible ρ-functions (the best among the worst).

6.4.7. Welsch's function

We now deal with another function of poor quality which looks like being a wild guess. It is introduced by Dennis and Welsch (1976) and obeys the equation

$$\rho(x-\vartheta) = \rho(u) = c^2 \left[1 - \exp{-(u/c)^2}\right].$$

(6.29)

The 0.95 asymptotic efficiency on the standard normal distribution is obtained with the tuning constant $c = 2.9846$.

6.4.8. "Bisquare" function

This function has been proposed by J. Tukey in the Princeton year on robustness. It is misleading due to its lack of unique solution and its particularly low β as defined at § 6.4.5. It has been specially investigated by Gross (1977) and corresponds with

$$\rho(x-\vartheta) = \rho(u) = \begin{cases} (c^2/3)(1-[1-(u/c)^2]^3), & \text{if } |u|/c \leq 1 \\ (c^2/3), & \text{if } |u|/c > 1. \end{cases}$$

(6.30)

The 0.95 asymptotic efficiency with respect to the standard normal distribution is obtained with the tuning constant $c = 4.6851$.

6.4.9. Andrews's function

As the previous function, this proposal has received a great deal of attention; but as people are more inclined to claim their successes than their failure, it still happens from time to time that a paper recommends the function. Very likely, Andrews has been a little too confident in his 1974-paper. It may even be thought that the data reported in the last columns of his Table 5 have never been computed; the last column is a copy of the last but one (although it should not be) and this last but one column is fairly inaccurate. This will be discussed at § 9.5. His function reads:

$$\rho(x - \vartheta) = \rho(u) = \begin{cases} 2c^2 \left[1 - \cos(u/c)\right], & \text{if } |u|/c \leq \pi \\ 4c^2, & \text{if } |u|/c \geq \pi . \end{cases} \tag{6.31}$$

The 0.95 asymptotic efficiency on the standard normal distribution is obtained with the tuning constant $c = 1.3387$.

6.4.10. Selection of the ρ-function

It seems difficult to recommend a ρ-function for general use without being rather arbitrary. Hence we will confine ourselves to comment upon the best candidates for the location (or regression) problems.

First we should note that quite many other proposals have been made. With regard to the location problem an impressive list of some seventy estimators have been tested on about two dozen distributions in the Princeton Monte Carlo project; the results are reported in Andrews et al (1972) and further details on the estimators are to be found in Gross and Tukey (1973). Some estimators involve several tuning constants, some of them being more or less adaptive and others being derived from highbrow concepts (e. g., Collins, 1976); we have a prejudice against estimators that are supported neither by theory, nor by any pragmatic feature (such a computational ease).

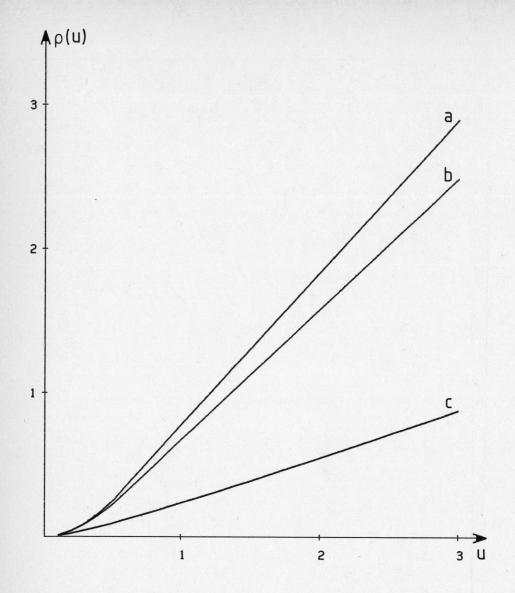

Fig.5. Comparison of the robust ρ-functions. (a) Function of Huber (6.25); (b) Modified Huber function (6.26); (c) Fair (6.27).

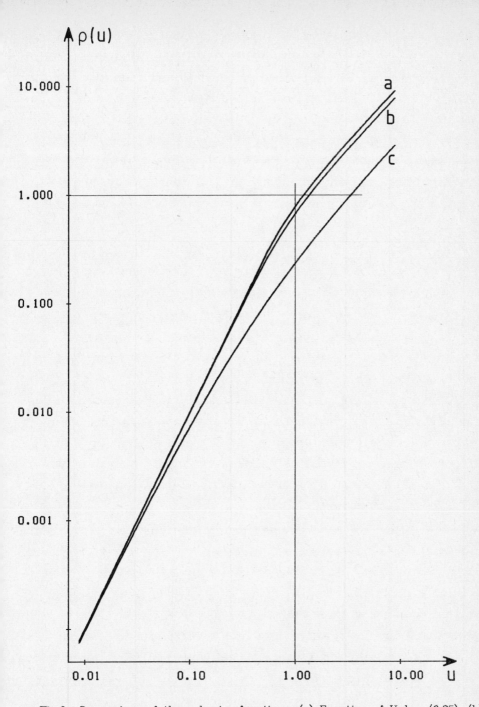

Fig.6. Comparison of the robust ρ-functions. (a) Function of Huber (6.25); (b) Modified Huber function (6.26); (c) Fair (6.27).

In order to recommend a ρ-function one has to keep in mind its use. Whenever a robust treatment of data is to be done with another more standard treatment, the bounding of the influence function is of no great importance; the major point is that the influence function must increase much more slowly than it does with standard methods. Then, Lp methods are the best in spite of their theoretical non-robustness: they are quasi-robust. The ρ-function (6.24) suffers from its computational difficulties but has the very great advantage of being scale invariant. Small values of the tuning constant ($\nu = 1.2$) should be used.

The second best ρ-function is without any doubt "Fair" by (6.27) because it hardly depends on the scale estimates. It is robust, has everywhere defined continuous derivatives of first three orders and as a consequence yields nicely converging computational procedures.

Eventually comes the Huber's proposal, either modified or in its original form. The modified form (6.26) is a little less sensitive to the scale estimate (with a larger β index) and benefits from a smooth second derivative; whereas Huber's proposal has the merit of being theoretically optimal in the asymptotic situation. With regard to computations, the modified form is to be preferred for algorithms of a general nature; however Huber's proposal (6.25) yields extremely fast algorithms when different treatments are applied to the observations according to their occurrence either in the strictly convex part or in the linear part of function (6.25).

A further view as to how the three functions Fair, Huber and modified Huber (6.25-27) behave can be derived from the two following displays. All functions have been multiplied by a factor such that they behave as u^2 for small variable u. Fig. 5 shows the three functions on linear scales whereas Fig. 6 permits assessment in a larger range thanks to the double-logarithmic scales. Each time the tuning constants have been adjusted to values that yield a 0.80-efficiency with respect to the standard normal law.

Chapter 7
Type L estimators

In this chapter we will be concerned with estimators that are "linear combinations of order statistics" in the strict sense of the words. A more general view will be given with the quantile estimators of Chap. 10, which are such that a given quantile of some error distribution is minimized.

7.1. Definition

To define L estimators without being too restrictive, we first consider a few specific cases.

In the standard one-dimension location-scale problem, the medians are frequently used as being unsensitive to outlying observations. They are so important that special attention is paid to them in § 7.3. A location estimator is given by

$$median \ (x_i)$$

and a scale measure by the median absolute deviation

$$MAD = median \ [\ |x_i - median \ (x_i)| \ ].$$

A more general structure for location estimators appears when comparing the two following proposals, the first is attributed to Edgeworth (1893) whereas the second is due to Tukey. Both are weighted averages of the first quartile, the median and the third quartile. Edgeworth's estimator is

$$(5 \, q_{1/4} + 6 \, q_{1/2} + 5 \, q_{3/4}) / 16$$

which is to be compared with Tukey's trimean

$$trimean = (q_{1/4} + 2 \, q_{1/2} + q_{3/4}) / 4.$$

In both cases, q_α stands for the empirical quantile of level α. A further variation may be obtained by considering other quantiles as well as another set of weights. For instance Gastwirth's proposal, which involves the one third and two thirds levels, is

$$(3\,q_{1/3} + 4\,q_{1/2} + 3\,q_{2/3}) \,/\, 10.$$

All these estimators are linear combinations of the order statistics of the available sample. The usual difficulties with defining the required quantiles are encountered but we will not deal with them here because the subject is well covered in books on order statistics as well as in works on non-parametric methods. A further reason for not going into this matter is the fact that the L estimators play a minor role in robust methods (except for the median); this is reflected in the little space devoted to them in the Huber's survey (1972), in Hogg's review (1979a) or in the simulation studies of Andrews et al (1972), and Rocke and Downs (1981). This view is contradicted by papers specially devoted to the one-dimension location problem such as the one of Chan and Rhodin (1980), in which many valuable references are to be found. David and Shu paper (1978) is also of interest in this connection.

The main difficulties are the limitation to one-dimension problems (this is relaxed in Chap. 10), how to cope with ties (see the fundamental study of Conover, 1973) and the low computational efficiency. This last point is the object of mixed feelings for it seems to be rather subjective; on the one hand rapid sorting algorithms are available and, on the other hand, approximate quantiles may be as good as exact quantiles (see von Weber, 1977).

Assume we have observed the sample $x_1,...,x_n$ at $p = 1$ dimension. According to (2.40) we may define the empirical density of probability

$$g\,(\,x\,) = (\,\textstyle\sum w_i)^{-1} \sum w_i\, \delta\,(\,x - x_i)\tag{7.1}$$

for any given set of positive weights $w_1,...,w_n$. To this density of probability corresponds a probability function which is uniquely defined by

$$G\,(\,x\,) = \int_{-\infty}^{x} g\,(\,x\,)\, dx;\tag{7.2}$$

this is a step function which leads to a difficult infinitesimal analysis. Therefore we render $G(x)$ in (7.2) everywhere continuous by substituting a kernel on non-zero width to each Dirac function in (7.1). For a positive infinitesimal η, we introduce an everywhere positive kernel of width η. The nature of the kernel does not matter; it may be a normal density with standard deviation η or a logistic density according to

$$\delta(u) = \frac{1}{\eta} \frac{\exp(u/\eta)}{[1 + \exp(u/\eta)]^2}. \tag{7.3}$$

This "trick" smoothes out the steps of $G(x)$, and yields a uniquely defined inverse; let it be noted $G^{-1}(\alpha)$. We have obtained a formal definition of the quantile q_α at level α. By definition this satisfies:

$$q_\alpha = G^{-1}(\alpha)$$

with

$$G[G^{-1}(\alpha)] = \alpha$$

and

$$G^{-1}[G(x)] = x. \tag{7.4}$$

With such a notation the trimean is defined as:

$$trimean = [G^{-1}(.25) + 2G^{-1}(.5) + G^{-1}(.75)]/4.$$

In general terms, an estimator of the type L is defined by the formula

$$\vartheta = \sum a_i \, G^{-1}[(i-1/2)/n] \tag{7.5}$$

where the coefficients a_i depend upon the sample size but are independent of the observations. They are usually associated with a score function J by

$$a_i = \int_{(i-1)/n}^{i/n} J(u) \, du. \tag{7.6}$$

119

It is appealing to combine (7.5) and (7.6) into

$$\vartheta = \int_0^1 J(u) \, G^{-1}(u) \, du, \tag{7.7}$$

but this is only correct in the equal-weight situation and under (7.3). However when quantiles of successive indices i are similarly weighted by coefficients a_i in (7.5), it might be inferred that $J(u)$ in (7.6) is rather smooth; then (7.7) closely approximates (7.5).

7.2. Influence function and variance

The discussion of the various steps leading to the influence function is rather tedious when the roughnesses of $G(x)$ by (7.2) and of $J(u)$ in (7.6) are taken into account. They will therefore be omitted and we refer to Boos (1979) as well as to Parr and Schucany (1982) for further details. We assume that both $G(x)$ and $J(u)$ have continuous first order derivatives everywhere defined; finally we relax the condition on $G(x)$.

The influence function will be worked out after its Gateaux's derivative definition (2.39). First, we introduce the perturbed probability density

$$g^*(x) = (1-t) \, g(x) + t \, \delta(x - x_0); \tag{7.8}$$

this problem yields an estimator

$$\vartheta^* = \int_0^1 J(u) \, (G^*)^{-1}(u) \, du. \tag{7.9}$$

Eventually, the influence function in x_0 can be expressed by:

$$\Psi(x_0) = \lim_{t \to 0} (\vartheta^* - \vartheta) / t. \tag{7.10}$$

$$= d\vartheta^* / dt.$$

We directly proceed under our assumptions.

$$\Psi(x_0) = (d/dt) \int_{-\infty}^{+\infty} J[G^*(x)] \, x \, g^*(x) \, dx$$

$$= \int_{-\infty}^{+\infty} [dJ(u) / du] \{ \int_{-\infty}^{x} [\delta(x - x_0) - g(x)] \, dx \} \, x \, g(x) \, dx$$

$$+ \int_{-\infty}^{+\infty} J\left[\, G\left(x\right)\,\right] x \left[\,\delta\left(x-x_0\right) -g\left(x\right)\,\right] dx$$

$$= - \int_{-\infty}^{+\infty} J\left[\, G\left(x\right)\,\right] \{ \int_{-\infty}^{z}\left[\,\delta\left(x-x_0\right) -g\left(x\right)\,\right] dx \} dx$$

The last expression results from integration by parts and is finite under very mild conditions. When both following integrals are finite, we may write

$$\Psi\left(x_0\right) = \int_{-\infty}^{x_0} J\left[\, G\left(x\right)\,\right] G\left(x\right) dx - \int_{x_0}^{\infty} J\left[\, G\left(x\right)\,\right]\left[\, 1-G\left(x\right)\,\right] dx \qquad (7.11)$$

Although we have assumed the differentiability of $G(x)$ in order to obtain (7.11), it is clear that we can relax that condition by considering some limiting scheme such as the one based on (7.3). It will be noted that our expression (7.11) does not concur with that proposed by Huber (1972, § 4.2) which is erroneous. Except for notational differences, we have obtained the equation (3.4) of Boos (1979).

For illustration, consider the evaluation of a quantile q_α as defined in (7.4). By (7.5-7), we see that it results from

$$J\left(u\right) = \delta\left(u-\alpha\right)$$

or equivalently

$$J\left(x\right) = J\left(u\right) / g\left(x\right)$$

$$= \delta\left(x-q_\alpha\right) / g\left(x\right).$$

This formula substituted in (7.11) yields the influence function

$$\Psi\left(x_0\right) = (\alpha-1) / g\left(q_\alpha\right), \qquad \text{if } x_0 < q_\alpha \qquad (7.12)$$

$$= \alpha / g\left(q_\alpha\right), \qquad \text{if } x_0 > q_\alpha.$$

A little less readily investigated situation is the case of the trimmed mean, an estimator which is claimed to be among the best robust candidates of location. The trimmed mean at level α is the mean of the remaining observations when a percentage α of the smallest and greatest observations have been discarded; thus

$$\vartheta = \textit{trimmed mean}$$

$$= (1-2\alpha)^{-1} \int_{q_\alpha}^{q_{1-\alpha}} x\, g\,(x)\, dx$$

$$= \int_0^1 J\,(u)\, G^{-1}\,(u)\, du \tag{7.13}$$

with

$$J\,(u) = 0, \qquad \text{if } u < \alpha.$$

$$= (1-2\alpha)^{-1}, \quad \text{if } \alpha \leq u \leq 1-\alpha.$$

$$= 0, \qquad \text{if } u > 1-\alpha.$$

The influence function immediately results from (7.11); it has a very simple form if the distribution is symmetric about some center m, namely

$$\Psi\,(x_0) = (q_\alpha - m)\, /\, (1-2\alpha), \quad \text{if } x_0 < q_\alpha$$

$$= (x_0 - m)\, /\, (1-2\alpha), \quad \text{if } q_\alpha \leq x_0 \leq q_{1-\alpha}$$

$$= (q_{1-\alpha} - m)\, /\, (1-2\alpha), \text{ if } x_0 > q_{1-\alpha}.$$

Using our previous derivations, we directly obtain a variance estimator from (4.31) and (4.35); it has the general form

$$Var\,(\vartheta) = [\, \textstyle\sum w_i /\, \bar{w} - 1\,]^{-1} \int [\, \Psi\,(x)\,]^2 g\,(x)\, dx \tag{7.14}$$

where, according to (4.37), the mean weight is

$$\bar{w} = \sum w_i^2\, /\, \sum w_i\,.$$

A very practical question we now must pose is whether an expression such as (7.14) has any meaning or not. The question may be approached from a few different angles of view; we feel that the root of the problem lies in what a variance (or a standard deviation) does mean. It is a way to assess a scatter of a random variate distributed (more or less) according to a normal law. But a look at (7.14) where $g\,(x)$ is the empirical distribution (7.1) reveals that the integral stands for a finite sum of components (7.11); this finite sum may yield a normally distributed result but that is only true if the components take many non-zero different values. Thus it may be expected that a reasonable estimator of the variance is obtained if, and only if, many quantiles are used or, in other words, if and only if $J[G(x)]$ is a smooth weight function. This is in essence what Mason (1981) as well as Parr and Schucany (1982) have observed.

We will not discuss the problems relative to robustness because they are very far from being resolved. Let us just note that $\Psi(x_0)$ has nice properties of smoothness (and boundedness) provided the weight function $J(u)$ is smooth and vanishes at the limits $u = 0$ and $u = 1$. This kind behavior of the influence function is favorable for robustness. How to design a weight function is discussed by Chan and Rhodin (1980); it will be seen that their technique is partially heuristic.

However it is appropriate to note that all comparison studies in which a trimmed mean (7.13) has been introduced into the set of tested location estimators have revealed a good or even outstanding performance for T_{10} or T_{15}, that is to say for the trimmed mean at the level $\alpha = .10$ or $\alpha = .15$. These estimators exhibit a satisfactory robustness and smooth influence functions; they cope with the most outlying observations without lowering the efficiency too much. The relative efficiency with respect to the normal distribution for a few specific values is listed in Table 7. It is of interest to compare the trimmed mean with the M estimators we have met. With respect to Huber's function (6.25), it can be seen that the essential difference is in the way of defining the "turning points"; in (6.25), the constant k defines these points to lie symmetrically with respect to the center whereas (7.13) does so with respect to quantiles. In some sense, it can be said that the function Fair (6.27) is a compromise in being very gradual in its turning points. Following Jewett and Ronner (1981), it must also be observed that the largest and smallest trimmed mean estimates bound the corresponding Huber's estimators. The feature of symmetry remains a difficult question whatever is the approach; it is not at all clear whether symmetrical distributions should be assumed for moderate size samples but, practically, there are few alternatives. Boyer and Kolson (1983) draw the attention on the selection of the trimming level: it should not depend on the observations when the sample size is moderate.

7.3. The median and related estimators

The median is an estimator which has attracted a great deal of attention for it is by nature very robust (we hesitate to say the "most" robust because of its possible lack of unicity). We have already seen that it can be much superior to the mean in assessing the location of contaminated normal distributions (see § 1.2). It is the only estimator presenting a breakdown point $\eta^* = 1/2$ (see § 3.2) and happens to be the limit solution obtained with the four robust ρ-functions (6.24-27). One of the reasons why potential users of the median estimator are discouraged is the fact that it is difficult to evaluate on a computer or, at least, relatively expensive in that it takes a long computer time. This view seems to be well-founded, however it may be worth-while to consider very fast algorithms such as those proposed by Schonhage, Paterson and Pippenger (1976) whenever large data sets are considered.

Before we proceed to discuss views which are more relevant to robust studies, let us remember that the median and the mean are not estimators of the same parameter except for symmetric distributions; however both estimators are location estimates and this is what matters. They have been compared in this aspect by Stavig and Gibbons (1977) who have attempted to obtain a sound basis of comparison; in fact it is not clear whether the variance is a sensible criterion although the tendency to approach normality is relatively rapid (see Wretman (1978) for an elegant derivation). A further item on the difference between the mean and the median is well known for moderately skew unimodal distributions: the median is located at one third from the mean in the direction to the mode. More generally the

Trimmed mean α	E Relative Efficiency	Turning point k in (6.25)
0.0500	0.9744	1.6448
0.0893	0.9500	1.3450
0.1000	0.9430	1.2816
0.1500	0.9092	1.0364
0.1631	0.9000	0.9818
0.2000	0.8736	0.8416
0.2322	0.8500	0.7317
0.2500	0.8367	0.6745
0.2983	0.8000	0.5294
0.3000	0.7987	0.5244

Table 7

interrelations between the mean, the median and the mode have been discussed by Runnenburg (1978) and van Zwet (1979) who present a list of references.

But should we use a median or a mean? As stated Kubat (1979), this is an old problem. We already saw in the conclusions of the Princeton Robustness study that the arithmetic mean was the worst estimator (see § 6.4.2), but this is opposed by Stigler (1977) who ranks the mean high and the median low. An intermediate view results from the investigations by Relles and Rogers (1977) and by Rocke, Downs and Rocke (1982). The root of the question obviously lies in the quality of the data set: for crude data the median is superior whereas for clean observations the mean is. However a consistent conclusion comes to light with respect to other estimators: expressed in the terms of Relles and Rogers "Use a popular robust estimator". This is noteworthy because many robust estimators use a median in some intermediate step of their definition (usually in a scale estimation). Moreover a median estimator is sometimes used to start an algorithm of robust estimation (see Harvey, 1977). It is rather exceptional to use a median as a final estimator except when it is justified by the field of application (e. g. Miller and Halpern, 1980); an interesting case is the multidimensional regression as considered by Cryer, Robertson, Wright and Casady (1972) who propose the median as the solution to a minimax problem.

Although the theoretical variance of the median is given by

$$Var \ (median) = [\ 4 \ (n-1) \ g(\ q_{1/2})^2 \ ]^{-1}$$

in accordance with (7.12) and (7.14) for a sample of size n, it is not clear what this variance is supposed to be in practical situations where the parent density of probability is unknown. A relatively natural way consists in estimating the density from the order statistics and this yields a kind of distribution-free estimated density. This approach, which is followed by Maritz and Jarrett (1978), has been supported by the Efron's bootstrap method. An approach such as the ordinary jack-knife applied to the median suffers from the fact that the definition of the empirical median differs for even and odd sample sizes. The infinitesimal jackknife is helpful but, then, the empirical density is estimated so locally that it is hardly reliable. Efron (1980b) has sur-veyed the results provided by the various similar techniques.

To conclude this chapter on L estimators, let us consider MAD, the median of the absolute deviations to the median, although it is a multi-ple L estimator rather than a simple L estimator. Its definition is such that it is optimally robust: its influence function is well controlled. As regards its nature, MAD is a natural candidate for estimating observa-tion scatter and is therefore frequently applied in robust estimation methods. Before we study an estimator family including MAD, let us mention its main properties; they immediately result from its definition

$$MAD = median \ [\ | x_i - median(x_i)| \ ] \, . \tag{7.15}$$

For symmetric (and moderately skew) distributions,

$$MAD = (q_{3/4} - q_{1/4})/2$$

and this displays its weak dependence on the median estimate, $q_{1/2}$. For normal distribution, we infer the standard deviation from.

$$\sigma \approx 1.4826 \; MAD \; .$$

The estimator MAD is a member of a much larger family of estimators based on folded distributions. Usually the folding takes place with respect to a location estimator ϑ_1 and the final estimator ϑ_2 is a scale estimate. Let us denote by $f(x)$ the distribution leading to ϑ_1, namely

$$\vartheta_1 = \vartheta_1 [f(x)] . \tag{7.16}$$

We use the symbol y to mean the absolute deviations to ϑ_1, and let $g(t)$ denote the corresponding distribution; i. e.,

$$y = |x - \vartheta_1|$$

$$g(y) = f(y - \vartheta_1) + f(\vartheta_1 - y) . \tag{7.17}$$

The estimator we are interested in is represented by ϑ_2 and can be derived from $g(y)$,

$$\vartheta_2 = \vartheta_2 [g(y)] . \tag{7.18}$$

As we know, the main properties of ϑ_2 are described by its influence function. It can be obtained by following the same approach as used before and is informally given below as in (7.8-10). First we perturb $f(x)$ into

$$f^*(x) = (1-t) f(x) + t \; \delta(x - x_0) ,$$

which induces a perturbation of ϑ_1 and $g(y)$ by (7.16-17)

$$\vartheta^*_1 = \vartheta_1 + t (d\vartheta^*_1 / dt)$$

$$g^*(y) = f^*(y - \vartheta^*_1) + f^*(\vartheta^*_1 - y)$$

$$= f^*(y - \vartheta_1) + f^*(\vartheta_1 - y)$$

$$+ t \; [f^{*'}(\vartheta_1 - y) - f^{*'}(y - \vartheta_1)] \; (d\vartheta^*_1 / dt)$$

$$=f^{*}(y-\vartheta_1) + f^{*}(\vartheta_1-y)$$

$$+ t\,[f'(\vartheta_1-y) - f'(y-\vartheta_1)]\,(d\vartheta_1^{*}/dt)\ .$$

Because the term between brackets cancels for symmetric distributions whenever ϑ_1 is an estimator of the center, we may think that the quality of ϑ_1 (robustness and scatter) does not matter too much. Specifically the median deviation to any robust location estimator is about as good as MAD can be. In what follows we will assume for simplicity that the distribution is symmetric and that a quantile of level α is to be estimated ($\alpha = 1/2$ would yield MAD). Finally we will study what quantile is the most appropriate to obtain a reliable estimate of the scatter.

In the present situation, the quantile of level α is ϑ_2 which satisfies

$$\alpha = \int_{0}^{\vartheta_2^{*}} g^{*}(y)\,dy\ .$$

Therefore the influence function is

$$\Psi(x_0) = (d/dt)\,\vartheta_2^{*} = \begin{cases} (\alpha-1)/g(\vartheta_2)\,, & \text{if } |x_0-\vartheta_1| < \vartheta_2\,, \\[2mm] \alpha/g(\vartheta_2)\,, & \text{if } |x_0-\vartheta_1| > \vartheta_2\,, \end{cases}$$

a result which resembles (7.12). In view of the symmetry we may write the variance in the form

$$Var\,(\vartheta_2) = \frac{1}{n-1}\alpha\,(1-\alpha)/[2f\,(\vartheta_1+\vartheta_2)]^2\ . \tag{7.19}$$

We observe that this variance may vary unsteadily for increasing parameter α; it is therefore natural to suspect the existence of an optimal value of α leading to a best scatter estimate. Let this best scatter estimate be that estimator which has the minimum variation coefficient; thus we search for α such that

$$Var\,(\vartheta_2)/\vartheta_2^2 \ be\ minimum$$

or, assuming $\vartheta_1 = 0$ without any loss of generality,

$$\alpha(1-\alpha)/[2xf(x)]^2 = \min \text{ for } \alpha \qquad (7.20)$$

under the constraint resulting from (7.17-18)

$$\alpha = 2 \int_0^x f(x)\, dx \ . \qquad (7.21)$$

We first verify our suspicion by demonstrating that (7.20) tends to infinity at the limits of the α-interval [0, 1] for most of the regular unimodal symmetrical distributions $f(x)$. For very small α, (7.21) yields

$$\alpha \approx 2\, x\, f(0)$$

and (7.20) behaves as

$$(1-\alpha)/\alpha \ .$$

For $\alpha \approx 1$, that is to say for large x, we may approximate the integral (7.21) with the help of

$$\int_x^\infty f(x)\, dx = -f^2(x)/f'(x)$$

as do Andrews (1973) and Gross and Hosmer (1978). This approximation implies an exponential behavior for the distribution tails and, then, (7.20) tends to infinity in accordance with

$$-\alpha/[2\, x^2\, f'(x)] \ .$$

The suspicion is further supported by numerical experiments. Table 8 contains the results obtained for a few distributions; it is seen that (7.20) is a relatively flat function and that the value $\alpha = 1/2$ of MAD is reasonable although certainly not the best. MAD has two great advantages: it has the best possible breakdown point and, furthermore, it is valued by many authors who put their "faith" in its questionable merits. Table 8 relates to the ratio of squared variation coefficients, a kind of relative efficiency. Let $\alpha^{\bullet}$ be the value of α such that (7.20) is a minimum, we then define

$$Relative\ efficiency = [(7.20) \text{ for } \alpha^{\bullet}]/[(7.20) \text{ for } \alpha] \ . \qquad (7.22)$$

Ratio of Squared Variation Coefficients

Distribution	α^*	Relative efficiency (7.22)			
		$\alpha = .5$	$\alpha = .6$	$\alpha = .7$	$\alpha = .8$
Cauchy	0.5254	0.9966	0.9708	0.8465	0.6461
Student, $\nu = 2$	0.6417	0.9064	0.9909	0.9803	0.8401
$\nu = 4$	0.7395	0.7718	0.9078	0.9910	0.9739
$\nu = 8$	0.7980	0.6769	0.8245	0.9455	1.0000
$\nu = 16$	0.8294	0.6225	0.7711	0.9056	0.9933
Normal	0.8617	0.5635	0.7093	0.8527	0.9694
Logistic	0.7941	0.6943	0.8385	0.9526	0.9998
Laplace	0.7968	0.7419	0.8643	0.9593	0.9999

Table 8

In view of the low efficiency of MAD, an estimator with $\alpha =. 7$ is recommendable, namely

$$Scatter \;=\; q_{.7} \; of \; \{|x_i - medium \; (x_i)|\} \; ;$$

such an estimator has similar performances with respect to long tail and normally distributed variates, moreover the breakdown point remains appreciable.

Chapter 8
Type R estimator

8.1. Definition

In the 1972 Wald lecture, Huber has been mainly concerned with robust location estimators. In order to present all possible estimator structures, he introduced three classes, M, L and R. We are now interested in the members of the third class, i. e. the estimators that have been obtained from Rank tests. They will be introduced according to Huber's viewpoint and therefore have little in common with the R estimators of Hogg (1979 a, b); the latter estimators can be regarded as an intelligent blend of M- and L estimators.

It may at first sight be surprising to observe that the very few papers treating R estimators have borrowed Huber's presentation (1972, 1977c). However there are two ingredients which can be separated at will. We first consider two-sample location tests and subsequently deal with one-sample rank tests.

Assume we have the disposal of two samples drawn from a common distribution except possibly for some slippage and let them be $X = \{x_1,...,x_n\}$ drawn from $f(x)$ as well as $Y = \{y_1,...,y_m\}$ drawn from $f(x+\vartheta)$. Then there exist a great many number of statistical techniques that can be used to assess whether the slippage parameter ϑ differs from zero, i. e. is positive or negative. All these techniques finally produce a test statistics t with the property of cancellating whenever the two samples nicely overlap. To put it briefly,

$$t = 0 \Longrightarrow overlap\ of\ X\ and\ Y. \tag{8.1}$$

Although the general theory has been established for samples drawn from identical distributions (except for the slippage), it is clear that (8.1) holds true even if the two distributions have different shapes. This is of importance because it is the only property we require of test statistics.

Now we are about to use the above slippage test dishonestly in order to define the R estimators. We suppose the sample $Z = \{z_1, \cdots, z_n\}$ to be located about some center ϑ and to be distributed according to $g(x+\vartheta)$ and derive from Z the following two overlapping samples:

$$X = \{x_1, \cdots, x_n\} = \{z_1-\vartheta, \cdots, z_n-\vartheta\} ,$$

$$Y = \{y_1, \cdots, y_n\} = \{\vartheta-z_1, \cdots, \vartheta-z_n\} .$$

Whereas X is distributed according to $g(x)$, the sample Y is distributed according to $g(-x)$. The two samples overlap only if the slippage parameter ϑ is appropriately defined and, reversing the logic of the slippage test, we define ϑ as being the parameter value such that the test statistics t cancels according to (8.1). In brief, ϑ is such that

$$t = t(\vartheta) = 0 \tag{8.2}$$

Huber has restricted himself to considering rank tests of the Wilcoxon type; to the best of the author's knowledge extension has not be studied. Moreover it may be inspiring to observe the similarity between (8.2) and (6.5) which relates to M estimation; in both cases ϑ is the parameter set such that an equation cancels. Certainly one merit of "rank" tests compared to all other tests is that they do not require any distribution assumption.

8.2. Influence function and variance

Except for minor details, the few derivations of the influence function are identical. We should like to recommend the nice paper by James and James (1979) who cover most of the theoretical aspects more thoroughly than Huber does; this paper is oriented towards a special field of application, the analysis of quantal bio-assay, but this calls for some further interest. A more general background information can be derived from Rousseeuw and Ronchetti (1979) or Lambert (1981) who investigate the sensitivity of tests. Ronchetti (1982) is more specially concerned by the development of optimally robust tests in the linear models.

We however keep the exposition of R estimators at the present level, on the one hand because of their limited applicability and, on the other hand, the little justification of the complex analytical procedure they involve. A relatively pragmatic view of the possibilities is given by McKean and Schrader (1980) (and their references) who examined estimation based on the distance between ranks rather than on the Euclidean distance.

Chapter 9
Type MM estimators

9.1. Definition

The *MM* estimators are generalizations of the type *M* estimators we have discussed in Chapter 6. The fact that they occur much more frequently than the simple *M* estimators is the main justification to treat them separately. It is especially the computational procedures that are usually oriented in such a way that an approximate factorization of the corresponding type *M* estimation problem takes place (see §9.8); this often permits the computer load to be reduced. We now introduce these *MM* estimators with frequent references to Chapter 6.

Consider some sample space, say $\Omega \subset R^p$, on which a density distribution $f(x)$, $x \in \Omega$, is defined; possibly this distribution is unknown except for some example $(x_1, \ldots, x_n)$ and then we take into account the empirical distribution (2.40), i. e.

$$f(x) = (1 / \sum w_i) \sum w_i \, \delta(x - x_i) \tag{9.1}$$

based on a set of non-negative weights $(w_1, \ldots, w_n)$. In this Chapter we investigate the estimation of a set of parameters $(\vartheta_1, \ldots, \vartheta_g)$ which are such that they minimize the functions $(M_1, \ldots, M_g)$, that is they are such that

$$M_j = \min \text{ for } \vartheta_j, \ j = 1, \cdots, g ,$$

where

$$M_j = \int \rho_j (x, \vartheta_1, \cdots, \vartheta_g) f(x) \, dx . \tag{9.2}$$

This definition makes explicit the identity of the parameters with their estimates.

The above structure is motivated by the frequent interdependence of various estimators. For instance, a scale estimator usually depends upon a previous estimation of some location estimators, e. g. the variance σ^2 may be defined by the M structure, for p = 1,

$$\int [(x-\mu)^2 - \sigma^2]^2 f(x)dx = \text{min for } \sigma^2 ,$$

where μ is the mean defined by another M-structure

$$\int (x-\mu)^2 f(x)dx = \text{min for } \mu .$$

These are in fact estimators of type multiple-M or, as we call them, MM estimators.

As we did previously, we assume that the functions M_j can be differentiated under the integral sign; then, and similarly with (6.5), the MM estimators can as well be defined by the following set of g equations

$$\int_\Omega \dot\rho(x, \vartheta_1, \cdots, \vartheta_g) f(x) \, dx = 0 . \tag{9.3}$$

It will be observed that a definition such as (9.3) does not imply anymore a positive definite matrix A by (6.6). The corresponding matrix A must only be non-singular for the solution $(\vartheta_1, \ldots, \vartheta_g)$; it has the form

$$A = \begin{pmatrix} A_{11} & \cdots & A_{1g} \\ \cdot & & \cdot \\ \cdot & & \cdot \\ \cdot & & \cdot \\ A_{g1} & \cdots & A_{gg} \end{pmatrix} \tag{9.4}$$

with

$$A_{jk} = \int_\Omega (\partial/\partial\vartheta_k) \dot\rho_j(x, \vartheta_1, \cdots, \vartheta_g) f(x) \, dx . \tag{9.5}$$

Usually the matrix A is not symmetric.

In our illustration, the set (9.3) has $g=2$ equations and is, for an empirical distribution $f(x)$,

$$\sum w_i \left[(x_i - \mu)^2 - \sigma^2 \right] = 0 \ ,$$

$$\sum w_i (x_i - \mu) = 0 \ .$$

Strictly speaking, μ is an M estimator, whereas σ^2 is an MM estimator. The corresponding matrix A is

$$A = \begin{bmatrix} -2 \sum w_i (x_i - \mu) & - \sum w_i \\ -\sum w_i & 0 \end{bmatrix}$$

In what follows, we will assume that the estimators ϑ_j are column-vectors (possibly of dimension 1). The comparison of the present derivations with the results on M estimators can partially be made by considering a vector ϑ having for components the column vectors ϑ_j

$$\vartheta = (\vartheta'_1, \cdots, \vartheta'_g)'$$

as well as by considering a minimization function such as

$$\rho(x, \vartheta) = \frac{1}{2} \sum_j [\dot{\rho}(x, \vartheta_1, ..., \vartheta_g)]^2 \ .$$

9.2. Influence function and variance

The influence function can be obtained by direct application of the Gateaux's derivative definition. First we perturb the density of probability $f(x)$ into

$$f^*(x) = (1-t) f(x) + t \ \delta(x - x_0)$$

and then expand the g expressions (9.3) with respect to the non-perturbed elements. We contract the notation and only take into account the first-order term in the perturbation, thus

$$\int \dot{\overset{\bullet}{\rho}}_j (x, \vartheta^*_1, \cdots, \vartheta^*_g) \, f^*(x) \, dx$$

$$= \int \dot{\overset{\bullet}{\rho}}_j \, f^*(x) \, dx$$

$$= \int \{ \dot{\rho}_j + \sum [(\partial / \partial \vartheta_k) \dot{\rho}_j] (\vartheta^*_k - \vartheta_k) \} \, [(1-t)f + t \delta(x_0)] \, dx$$

$$= \int \dot{\rho}_j \, f \, dx + (1-t) \sum A_{jk} \, (\vartheta^{*}_{k} - \vartheta_k) + t \, \dot{\rho}_j(x_0)$$

$$= \sum A_{jk} \, (\vartheta^{*}_{k} - \vartheta_k) + t \, \dot{\rho}(x_0) = 0 \ .$$

We see that the difference $(\vartheta^{*}_{j} - \vartheta_j)$ is the solution of a set of g linear equations. They may be scalar but sometimes they are vectorial or functional, depending upon the nature of the estimators. When the coefficients A_{jk} $(k \neq j)$ are dominated by A_{jj}, that is when the estimators are relatively independent of one another, an explicit solution can be derived. Under the condition

$$|| \sum_k A_{ik} \, A_{kk}^{-1} \, A_{kj} \, A_{jj}^{-1} || << 1 , \quad k \neq i , k \neq j , \tag{9.6}$$

we obtain the influence functions

$$\Psi_j(x_0) = \lim_{t \to 0} (\vartheta^{*}_{j} - \vartheta_j) / t$$

$$= A_{jj}^{-1} \, [\sum_{k=1}^{g} A_{jk} \, A_{kk}^{-1} \, \dot{\rho} \, (x_0) - 2 \dot{\rho}_j(x_0)] \tag{9.7}$$

In this result we recognize a component-wise approximation of (6.13). The explicit writing of the influence function by (9.7) is convenient, although not necessary. When we relax the condition (9.6) on the matricial norms, we can directly solve the equation

$$\begin{bmatrix} A_{11} & \cdots & A_{1g} \\ \cdot & & \cdot \\ \cdot & & \cdot \\ \cdot & & \cdot \\ A_{g1} & \cdots & A_{gg} \end{bmatrix} \begin{bmatrix} \Psi_1(x_0) \\ \cdot \\ \cdot \\ \cdot \\ \Psi_g(x_0) \end{bmatrix} + \begin{bmatrix} \dot{\rho}(x_0) \\ \cdot \\ \cdot \\ \cdot \\ \dot{\rho}_g(x_0) \end{bmatrix} = 0 \ .$$

The variance-covariance matrix of an MM estimator is immediately derived from (4.31) and (4.35). Here it is written for an equally weighted sample

$$Cov \, (\vartheta_j) = \frac{1}{n(n-1)} \sum [\Psi_j(x_i)] \, [\Psi_j(x_i)] \, ' \tag{9.8}$$

and corresponds with (6.14).

9.3. Linear model and robustness - Generalities

Prior to studying the very frequent problems of robust location and robust regression in the next section, it might be appropriate to recall a few items of the standard linear theory and compare them with what our past derivations suggest in these situations. In fact most of this section could be transferred to Chapter 6 for we shall concern ourselves with the ordinary least squares, that is to say with simple M estimators.

First we note that we have to avoid any assumption on a normal distribution of error terms; this point must be stressed. Effectively it is not consistent to assume normally distributed errors and to consider robust techniques. If such an assumption were correct, then standard methods would be the most appropriate.

Rather than the classical writing of the linear model

$$y = x\beta + \varepsilon ,$$

we use the notation

$$\underset{(n\times 1)}{y} = \underset{(n\times g)}{z} \ \underset{(g\times 1)}{\vartheta} + \underset{(n\times 1)}{\varepsilon} . \tag{9.9}$$

The set of parameters ϑ (unlike that of β) should remind us that we should not imply the normality assumptions. The matrix z in (9.9) is introduced to be consistent with our previous notation where an observation is described by the vector x_i. In this context, an observation is a combination of one y-component and one z-row, i. e.

$$x_i = (y_i, z'_i)' . \tag{9.10}$$

The error vector ε results from observation errors; an error ε_i is associated with the observation x_i without any particular association with the first component of x_i. Thus we adopt a context of linear model with errors in the variables -See Fuller (1980), Gaber and Klepper (1980), Gleser (1981) as well as Golub and van Loan (1980) for recent literature- However, as we do not separate the error components and do not explicitly take the covariance between these components into account, our derivation remains very pragmatic. The only essential hypothesis is with respect to the nature of the errors: they are assumed to be random rather than systematic, in other terms they have a zero mean. All the other assumptions result from the estimation rule; we define $\hat{\vartheta}$ on the basis of the least squares criterion

$$M = \sum \hat{\varepsilon}_i^2 = \min \text{ for } \hat{\vartheta}$$

or explicitly

$$M = [y - Z\hat{\vartheta}]'[y - Z\hat{\vartheta}] = \min \text{ for } \hat{\vartheta} . \tag{9.10}$$

Using $E\{.\}$ to denote the expectation on the random components, we obtain $\hat{\vartheta}$ by differentiation of (9.10) and therefrom the main properties of the estimator. The least squares criterion leads to

$$\hat{\vartheta} = (Z'Z)^{-1} Z'y \tag{9.11}$$

under the implicit assumption that z is full rank. From (9.9), we have

$$\hat{\vartheta} - \vartheta = [(Z'Z)^{-1} Z'(Z\vartheta + \varepsilon)] - \vartheta$$

$$= (Z'Z)^{-1} Z'\varepsilon$$

as well as

$$\hat{\varepsilon} = y - Z\hat{\vartheta}$$

$$= [Y - Z(Z'Z)^{-1} Z']\varepsilon ,$$

and the estimator is unbiased if (and only if) the errors do not depend upon z

$$E\{\hat{\vartheta}\} = \vartheta + E\{(Z'Z)^{-1} Z\varepsilon\} = \vartheta . \tag{9.12}$$

Analogous to a mean-square-error rather than to a variance, we evaluate the second order moment

$$Cov(\hat{\vartheta}) = E\{(\hat{\vartheta} - \vartheta)(\hat{\vartheta} - \vartheta)'\}$$

$$= E\{(Z'Z)^{-1} Z' \, \varepsilon \, \varepsilon' \, Z(Z'Z)^{-1}\} \; ; \qquad (9.13)$$

in which any possible bias has been incorporated. Before proceeding let us return to the least squares criterion. It is usual to associate a variance measure with its value according to

$$n \, \hat{\sigma}^2 = \varepsilon' \varepsilon$$

or, at the minimum of (9.10), according to

$$M = \hat{\varepsilon}' \, \hat{\varepsilon}$$

$$= \varepsilon' \, [I - Z(Z'Z)^{-1} \, Z']^2 \, \varepsilon$$

$$= \frac{n-g}{n} \varepsilon' \varepsilon \; ;$$

from which we derive

$$\hat{\sigma}^2 = \frac{1}{n-g} M \, . \qquad (9.14)$$

which will be generalized in the next section in order to assess the mean scatter of the error components.

At this point, we have to compare the exact second order moment (9.13) with the result (6.14) we obtained from theory. In the present derivation all the observations x_i are equally weighted, say with

$$w_i = \bar{w} = 1 \, ,$$

and, according to (9.10) and (6.1), the ρ-function is

$$\rho(x_i, \vartheta) = (y_i - z'_i \vartheta)^2$$

From this we can deduce

$$\dot{\rho}(x_i, \vartheta) = -2(y_i - z'_i \vartheta) \, z_i$$

$$\ddot{\rho}(x_i, \vartheta) = 2 \, z_i z'_i$$

$$A = \frac{2}{n} \sum z_i z'_i = \frac{2}{n} Z'Z$$

$$B = \frac{4}{n} \sum (y_i - z'_i \vartheta)^2 z_i z'_i$$

$$= \frac{4}{n} \sum \hat{\varepsilon}_i^2 z_i z'_i$$

and finally

$$Cov(\hat{\vartheta}) = \frac{n}{n-1} (Z'Z)^{-1} (\sum \hat{\varepsilon}_i^2 z_i z'_i) (Z'Z)^{-1} \tag{9.15}$$

which should estimate the exact (9.13). We observe that these two formulae differ in their central parts, although their difference is much smaller than it might seem at first sight. Indeed, if we expand the central part of (9.13)

$$Z' \varepsilon \varepsilon' Z = \sum_i \sum_j \varepsilon_i z_i z'_j \varepsilon_j$$

$$= \sum_i \varepsilon_i^2 z_i z'_i + 2 \sum_{i>j} \varepsilon_i \varepsilon_j z_i z'_j \; ,$$

we see that (9.15) is a reasonable estimator whenever

$$\varepsilon_i z_i \text{ and } \varepsilon_j z_j \text{ are independent} \tag{9.16}$$

or, as is often the case in practice, are moderately correlated. In fact, (9.16) corresponds exactly to (4.14). The condition (9.16) is satisfied in most cases, even in the errors-in-variables linear models where ε_i and z_i are essentially interdependent.

That the estimator (9.15) of the variance-covariance matrix incorporates a possible bias is seen by the author as a very favorable feature.

With regard to robustness, it is clear that the least square criterion cannot yield robust estimators. A single outlying x_i can offset the estimate beyond any limit or, in other words, the breakdown point of the estimator is zero. An interesting pont is the combined role of ε_i and z_i; it not only tolerates errors in the variables or, as some say, errors in the carriers and leverage unbalance, but it also permits model inadequacies. For illustration, suppose that the linear model is in fact a first approximation to a quadratic model. Then condition (9.16) is not strictly satisfied, nevertheless the covariance between the $\varepsilon_i z_i$ and $\varepsilon_j z_j$ tends to cancel.

In order to robustify the estimates, we should bound the ρ-function and discard the least squares approach. As we still look for estimation rules which yield small residuals, we may write the estimating procedure in the form

$$M = \sum \rho(x_i,\vartheta) = \sum \rho(\hat{\varepsilon}_i) = \min \text{ for } \vartheta \ . \tag{9.17}$$

This leads to an estimator which can be seen as a function of y_i and z_i in model (9.9), say

$$\hat{\vartheta} = \hat{\vartheta}(y,Z)$$

or, stressing the error term,

$$\hat{\vartheta} = \hat{\vartheta}(\vartheta,Z,\varepsilon) \ .$$

A natural requirement is that $\hat{\vartheta}$ be "scale invariant" or, in other words, we are interested in a family of estimators which do not depend on the parameter λ in

$$\hat{\vartheta} = \hat{\vartheta}(\vartheta,Z,\lambda\varepsilon) \ .$$

It happens (Rey, 1978, p. 43) that the only possible ρ-function in (9.17) is the power function. This is very inconvenient because the power functions do not yield robustness. The way out of this difficulty is to scale the residuals in order to obtain scale invariance. The solution is in the system of two equations

$$M_1 = \sum \rho(x_i,\vartheta) = \sum \rho(\hat{\varepsilon}_i/s) = \min \text{ for } \vartheta$$

$$s = scale \ of \ \varepsilon_i \ . \tag{9.18}$$

When the scale equation is an M estimator we have in fact a set of MM estimators. Because this question has the utmost importance, it will be treated in greater detail in the next section.

The linear set up has been very thoroughly investigated by many investigators, because it lends itself particularly well for adequate comparisons and easy simulations.

9.4. Scale of residuals

As expressed by (9.18), the robust estimation of the linear model (9.9) implies an estimation of the scale of the error components ε_i. What we mean by scale can remain rather vague inasmuch as it is an item with constant proportionality to the scatter, the standard deviation, the range or any other spread measure of the error components. For convenience, the scale estimator is usually standardized in such a way that it corresponds with the standard deviation whenever the variates are normally distributed.

But what is the "scale" of the residuals? A partially disappointing answer to this has been proposed by Huber (1964). We excerpt;

> The theory of estimating a scale parameter is less satisfactory than that of estimating a location parameter. Perhaps the main source of trouble is that there is no natural "canonical" parameter to be estimated. In the case of a location parameter, it was convenient to restrict attention to symmetric distributions; then there is a natural location parameter, namely the location of the center of symmetry, and we could separate difficulties by optimizing the estimator for symmetric distributions (where we know what we are estimating) and then investigate the properties of this optimal estimator for nonstandard conditions, e. g., for nonsymmetric distributions. In the case of scale parameters, we meet, typically, highly asymmetric distributions, and the above device to ensure unicity of the parameter to be estimated fails. Moreover, it becomes questionable, whether one should minimize bias or variance of the estimator.

Most authors have turned the above difficulties in two ways; first a scale parameter being (arbitrarily) selected, it is demonstrated that the selected scale equates the standard deviation under asymptotic conditions, secondly and pragmatically, the above authors tend to ignore the importance of the selection. A notable exception to these attitudes appears in the paper of Hill and Holland (1977) where the use of MAD, the Median Absolute Deviation, is questioned. They even conclude that the scale definition is of the utmost importance for the quality the estimation.

We have already met with quite a few estimators which are appropriate, or can be contrived so as to be appropriate, for scale estimation. The most frequent ones, at least for robustness studied, are L estimators and specially MAD by (7.15). We have seen that these estimators are robust but do not have a smooth influence function. This lack of smoothness may degrade seriously the performance of a given M estimator and prohibits any covariance estimation by formula (9.8). Let us discuss for a while these remarks to emphasize their importance.

An illustration is all we need. Let us consider a simple one-dimension location problem with MAD as scale estimator. We estimate the location of the sample $\{x_1, \ldots, x_n\}$ by the estimator $\hat{\vartheta}$ satisfying

$$M = \sum \rho \left[(x_i - \vartheta)/s \right] = \min \text{ for } \vartheta$$

with

$$s = MAD = median \{|x_i - \vartheta|\}$$

Rather than MAD, a multiple of MAD or another L estimator could be considered, but this does not matter. The ρ-function can be specified at will among the proposals (6.25 - 31) and this does not matter either. Interesting is the variation of MAD seen as a function of a varying parameter ϑ. For simplicity we will assume an odd sample size n. Then, for very large negative ϑ, we have

$$s = median \{x_i\} + |\vartheta| \ .$$

$$= x_{(m)} + |\vartheta|$$

$$= |x_{(m)} - \vartheta|$$

We now gradually increase the ϑ-value and see that the formula remains valid (with s decreasing) until

$$|x_{(m)} - \vartheta| = |x_{(1)} - \vartheta|$$

or

$$\vartheta = [x_{(1)} + x_{(m)}]/2 \ .$$

Now continuing to increase ϑ, we have

$$s = |x_{(1)} - \vartheta|$$

and this is valid until it is replaced by

$$s = |x_{(m+1)} - \vartheta| ;$$

finally we obtain for large positive ϑ-value

$$s = |x_{(m)} - \vartheta| .$$

the formula we had for large negative ϑ-values. Altogether we see that s is a piecewise linear function of ϑ with the slopes

$$ds/d\vartheta = +1 \text{ or } -1 ;$$

if the sample size had been even we would have obtained a piecewise linear function with the slopes

$$ds/d\vartheta = +1, 0, \text{ or } -1 .$$

Thus, we have seen that s is a continuous function of ϑ with a bounded first derivative fluctuating between two positive and negative bounds (this conclusion also holds good in regression analysis). Let us now return to our first equation. We define $\hat{\vartheta}$ as minimizing the sum

$$M(\vartheta) = \sum \rho \left[(x_i - \vartheta)/s \right] ,$$

and consider the values taken for increasing ϑ. The shaky behavior of s can easily induce shaky variations of the summed components and, finally, we observe a function $M(\vartheta)$ with several minima. Apparently, there would be several solutions to our location problem as to the values of $\hat{\vartheta}$ and this is a situation we cannot tolerate.

The only possibility to avoid the above situation is to define a scale estimator which has a single minimum and varies monotonously with increasing distances to this minimum . This orientation has been investigated without much success by Jaekel (1972) for L estimators; he even used the scale itself as the criterion for minimization and,

145

although he was able to cope with the problem of non-unique discrete solutions, he felt uneasy when dealing with solutions covering an interval or a domain.

Quite a different issue is the estimation of the variance-covariance matrix of the estimate. It turns out that the oscillating slope of s with respect to ϑ induces a fairly unstable matrix component A_{jk} by (9.5) and finally a fairly unreliable estimate results from (9.8). This can also be partially attributed to the moderate efficiency of the scale estimate and we have seen in Chapter 7 that the efficiency could be low for L estimators. This question has been reviewed by Fenstad, Kjaernes and Walloe (1980), with the support of simulation experiments, as well as by Johnson, McGuire and Milliken (1978).

But let us be consistent? What should we do?

The main idea in estimation theory consists in defining an estimator as a parameter set that yields small residuals.

Estimate <==> small residuals .

But what do we mean by "small" residuals? In fact we mean the "smallest" residuals, i. e. the smallest residuals measured in any suitable measure or, equivalently, the smallest measure of residuals. Therefore we may write

Estimate <==> smallest measure of residuals .

We have a little progressed, but it remains for us to define how to measure the residuals. They are positive or negative, and can even be vector-valued and it seems that a suitable measure is the space they cover or, in other words, their scatter. Thus we have

Estimate <==> smallest scatter of residuals .   (9.19)

In our present context of M estimation, we have defined $\hat{\vartheta}$ as the parameter set that minimizes

$$M(\vartheta) = \sum \rho(x_i, \vartheta)$$

and, therefore, $M(\vartheta)$ is a measure of scatter except for some monotonous transformation. Moreover, what transformation this is may be very clear. For instance, we have seen that the least squares procedures in the linear set-up imply

$$s^2 = \frac{1}{n-g} M = \frac{1}{n-g} \sum \rho(x_i, \vartheta)$$

More generally, since the ρ-functions (6.24-31) have a tendency to down-weight the large residuals it may be expected that such a formula tends to underestimate the squared scale. This leads us to define the scale estimator as:

$$s^2 = \frac{1}{(n-g)\, Int} \sum \rho(x_i, \vartheta) \tag{9.20}$$

where the coefficient Int is inferior or equal to 1, the value taken for the least squares as in (9.14). The appropriate value is defined in order not to introduce any bias on normally distributed variables, i. e.

$$Int = \int_{-\infty}^{+\infty} \rho(u)\, n(u)\, du \tag{9.21}$$

with

$$n(u) = (2\pi)^{-1/2} \exp(-u^2/2).$$

Let us note that the values which are taken on by the integral have been reported for the eight functions (6.24-31) in the last column of Table 6.

A very interesting by-product of the scale definition (9.20) is a simplification of the influence function equation (9.7). Let us write our estimation problem according to the general presentation (9.2-3); the estimator ϑ will be ϑ_1, while ϑ_2 stands for s^2, the squared scale. We have

$$M_1 = \sum \rho_1(x_i, \vartheta_1, \vartheta_2) = \left[\sum \rho(\hat{\varepsilon}_i / \sqrt{\vartheta_2})\right] \vartheta_2$$

and

$$\partial M_1 / \partial \vartheta_1 = \sum \dot{\rho}_1(x_i, \vartheta_1, \vartheta_2) = 0$$

as well as

$$dM_2/\,d\vartheta_2 \;=\; \sum \dot{\rho}_2(x_i,\vartheta_1,\vartheta_2)$$

$$=\; \sum [\vartheta_2 - K\rho_1(x_i,\vartheta_1,\vartheta_2)] \;=\; 0$$

where K is a data-independent constant

$$K \;=\; n/\,[(n-g)\,\mathit{Int}\,] \tag{9.22}$$

We know proceed to determine A_{21} by (9.25) and note in passing that the scale is defined by a good M estimator which is differentiable beyond any criticism whenever the criterion M_1 is differentiable. We obtain

$$A_{21} \;=\; (\partial/\,\partial\vartheta_1)\,dM_2/\,d\vartheta_2$$

$$=\; -K\sum \dot{\rho}_1\,(x_i,\vartheta_1,\vartheta_2)$$

$$=\; 0\;.$$

This result yields the verification of the condition (9.6) on matrix norms and simplifications in (9.7). The results are now exact rather than approximate; the two components of the influence function are

$$\Psi_2(x_0) \;=\; -A_{22}^{-1}\,\dot{\rho}_2\,(x_0,\vartheta_1,\vartheta_2)$$

and

$$\Psi_1(x_0) \;=\; -A_{11}^{-1}\,[\,\dot{\rho}(x_0,\vartheta_1,\vartheta_2) + A_{12}\,\Psi_2(x_0)\,]\;. \tag{9.23}$$

The squared scale estimator $\vartheta_2 = s^2$ behaves exactly as a simple M estimator (see 6.13). The corresponding ρ-function is $\rho_2\,(x_i,\vartheta_1,\vartheta_2)$ which has to be considered in terms of varying ϑ_2; in these terms it can hardly be robust. Therefore the influence function $\Psi_2(x_0)$ is usually not bounded and $\Psi_1(x_0)$ cannot be either. About the most we can say is that (9.20) is less sensitive to possible outliers than the squared residuals would be. In agreement with § 6.4, we have defined a quasi-robust estimator ϑ_1 whenever the minimized criterion is robust.

It is clear that we recommend to evaluate the scale by (9.20); this establishes the minimized criterion as a measure of the squared scale. Other proposals based on relations between the criterion and the scale have been put forth but, except one, none has retained the interest. Huber and Dutter (1974) as well as Huber (1977c) propose to minimize another expression than $(M_1.M_2)$ such that they obtain ϑ_1 and ϑ_2. Their expression, solved by Dutter (1977),

$$k \sum \rho(x_i, \vartheta, s^2)/s + s = \min \text{ for } \vartheta, \quad s > 0$$

does not seem very appealing to us in spite of its favorable properties; we do not like its high level of arbitrariness. The constant k is data-independent; it is defined similarly to our parameter κ by (9.22) or by a very much corresponding formula.

9.5. Robust linear regression

This is a widely investigated problem; it is sufficiently simple to permit a clear problem statement and sufficiently involved to entail endless discussions on its solutions. These features indicate that robust linear regression is a "good" topic of research; furthermore its field of application is very large but most authors seem to disregard this.

There are relatively few survey papers and most of them are derived (at least in part) from Huber's Wald Lectures (1973). We would like however to recommend those statisticians who have a knowledge of French to read the paper of Morineau (1978); it is written at an introductory level and clearly points out the practical aspects. Quite a few papers only deal with interesting theoretical aspects but remain at the level of asymptotic properties and, for that reason, are not referred to in what follows. First we mention the fundamental contribution of Ghosh and Sinha (1980) on the various relationships between the ordinary least squares, the maximum likelihood approach and robust techniques in the multivariate normal regression model and more or less non-normal error models; this paper is on the verge of establishing a clear-cut theoretical set-up encompassing most of the conceptual difficulties, in spite of its much more limited objectives. The less-covered topics are exposed by Bickel (1978) and Beran (1982) who have investigated heteroscedastic regression models (non-equal variances) and model inadequacies such as light non-linearity. The same topics have been formalized by Carroll and Ruppert (1979), Ruppert and Carroll (1979 b, c) and Carroll (1980 b) in a series of memoranda. We have considered these questions in the context of the function Fair (6.27). A related question appears with regard to errors in the variables; it has already been treated in § 9.3 and we will soon return to it. In this respect Hill's thesis (1977) has attracted much attention; the works of Maronna, Bustos and Yohai (1979), Yohai and Maronna (1979), Hill (1979) and most certainly of Krasker (1980) deserve special mention.

As we have already indicated, the robust solution of model (9.9) with a scale given by (9.20) results from the system

$$s^2 \sum \rho \left[\left(y_i - z'_i \vartheta \right) / s \right] = \min \text{ for } \vartheta,$$

$$1 = \frac{1}{(n-g) \, Int} \sum \rho \left[\left(y_i - z'_i \vartheta \right) / s \right], \tag{9.24}$$

the variates $[(y_i - z'_i) / s]$ being scale independent. How to solve such a system will be treated in § 9.8. The selection of the ρ-function has been discussed and we recommend the function Fair (6.27) whenever heteroscedasticity or errors on the z components are likely to occur.

With regard to the robustness of the estimate $\hat{\vartheta}$, we see that, by applying (9.23) and neglecting the term in A_{12}, the influence function at observation x_i

$$\Psi (x_i) = \{ \frac{1}{n} \sum [(d^2 / du^2) \rho (u)] z_j \, z'_j \}^{-1} s \left[(d / du) \rho (u) \right] z_i \tag{9.25}$$

is proportional to the regressor vector z_i and this feature occurs whatever has been the ρ-function selection. The situation in which the term in A_{12} cannot be neglected will be dealt with at § 9.7. Therefore robustness cannot be achieved with respect to errors in the variables. This situation can be handled at the cost of different ρ-functions for different z_i; in other words we need ρ-functions which are sensitive to y_i

and z_i rather than to the resulting residual. This possibility has been suggested in § 6.3 and some of the very intricate difficulties are faced by Velleman and Ypelaar (1980). An excellent paper of Welsch (1979) surveys the recent developments in the relevant asymptotic theories as well on the various proposals for bounding the influence function; it introduces to the views defended by Krasker and Welsch (1982) and Huber (1983).

It seems that very few authors have attempted to test robust methods by Monte Carlo simulation. In fact quite a few papers deal with location estimation (see next section) but hardly any general simulation has been attempted for solving serious regression problems. Limited results have been reported by Ramsay (1977) and their relevance is not clear to the author; we are left with the vague feeling that the paper aims at demonstrating the quality of proposed ρ-functions rather than assessing when one estimator outperforms another. In contrast, Hill and Holland (1977) are very open in comparing the least squares estimator against the least absolute power (6.24, $\nu = 1$) and Andrews' proposal (6.31). The least squares estimator is dominated by the two robust proposals whatever are the design matrix z and the contamination level of the error distribution. We should possibly also mention simulation studies made in order to test what structure is appropriate for the estimate variance-covariance matrix. In this regard the conclusions are so limited that we prefer to ignore the subject. However it might be interesting to recommend Hill's approach (1979) for future investigation. Most variance-covariance matrices appear to be rather arbitrary generalizations of the ordinary theory; they assume moderate normality and independency but, then, do we need robust methods? To the best of our knowledge, the incidence of the scale estimation (see factor A_{12} in 9.23) has never been noticed.

A common feature of most simulation studies is the difficulty to compare estimators with a standard solution, because such a standard solution does not exist in most cases. For instance, there is no sense in comparing with a "theoretical" solution what produce M estimators; the theoretical solution being derived under assumptions, one or the other estimator will be systematically better according to the types of departure from assumptions. We believe that equal efficiencies should be achieved by all the estimators compared. It does not make sense to compare estimators which are investing different powers on being robust; we feel concerned by the balance between efficiency loss and achieved robustness.

Because of the difficulty of undertaking an adequate simulation study, it may be appropriate to compare the various methods we have encountered by considering a well investigated case. We have selected the famous stackloss data set relative to the operation of a plant for the oxidation of ammonia to form nitric acid presented by Brownlee (1965, § 13.12). It is a linear regression problem of size $n = 21$ with $g = 4$ regressors, one of them being the unit constant. Not much physical information about the model is available; the variates are different in nature (see the references) and the linearity is questionable. The statistical information is also rather scanty. There are essentially two points recommending this data set; it is a set of real data and this set has been examined without too many prejudices as to what it contains. However, as the number of studies increased, ideas on what the set may contain developed and as also prejudices as to what should be contained in it and in the results of the analysis. However we do not ourselves, expect to be free from prejudices.

Let us review in turn a few steps in the various studies. Thus, in 1965, Brownlee incorporates the data set in a revision of his book of 1961. What (linear) model is appropriate has been extensively tested by Draper and Smith (1966, Ch. 6) with the help of many computer runs and plot assessments. Then, Daniel and Wood (1971, Ch. 5) observing a specially large fit residual for observation 21, have initiated a course of actions which seems flourishing; an extensive study led them to conclude that observations 1, 3, 4 and 21 were outliers. The robustness era starts with this analysis. Andrews (1974) presents his function (6.31) and tests its performance on the data set, however for reasons we do not understand he produces fake results - Possibly he was too much confident in the outcome - . The true results almost resemble the residuals obtained when fitting them to the 17 "normal" observations but are different at the second decimal place; they are reported in Rey (1977). We were privileged to receive a letter from D. F. Andrews (on Nov. 21, 76) in which he supports our findings. Andrews' paper confirms the existence of the 4 outliers. A little later a memorandum of Lenth (1976) reports a run of Andrews' method after initialization with Huber's method; he concludes that observation 13 is outlying as well. Based on the ranks of the residuals, a method of Hettmansperger and Mc Kean (1977) clearly points out that the four observations 1, 3, 4 and 21 are outliers; observation 2 is also outlying but with a negligible residual which is attributed to the structure of the design matrix. Plots may be very useful for identifying critical observations; a nice example is given by Denby and Mallows (1977) and clearly the four "outliers" behave strangely whereas observations 2 and 13 appear to be normal ones (see B and M in their Fig. 1). Observation 2 is criticized by Andrews and Pregibon (1978) who find five outliers. Aitkin and Wilson (1980) have used a missing value method to check whether an observation is outlying; considering how much computer time they required, we have the feeling that a robust method (whatever slow it may be) might be preferred. An adaptive procedure selecting among robust methods is proposed by Moberg, Ramberg and Randles (1980); they are fully convinced that the data set include four outliers and use for comparison the results obtained from an ordinary least squares method applied to the remaining 17 observations. Among their results, they report the erroneous Andrews' displays (main errors are inaccuracies in the second decimal place and the permutation of two residuals, $\hat{\varepsilon}_5 =$ -. 719 and $\varepsilon_6 = -1.238$). The faith in the existence of four outliers is a little open to question while reading Atkinson (1980) :

> The analysis in which the plots (giving confidence intervals) were employed together with power transformations led to a model for all 21 observations in which there was no evidence of outliers.

The same author indicates in 1981 that when observation 21 is rejected the residuals nicely fit in their respective confidence intervals; further rejection of observations 1, 3 and 4 also gives a good plot but the last three observations have to be rejected at the same time. Chambers and Heathcote (1981), with an original adaptive method for analyzing the variance of the residuals, confirm the four observations in their status of outliers; they also demonstrate that the remaining 17 observations nicely behave as if they were drawn from a multivariate normal distribution. To conclude this long list of findings, we may say that four observations are clearly abnormal with respect to the linear regression modeling; they are outliers in this context, but whether this is due to errors on the dependent variates or errors on the predictors will never be known.

Before we discuss the results obtained by quasi-robust methods while processing Brownlee's stack loss data, let us draw the attention to the fact that differences occur already at the level of the least squares treatment. In the present situation it does not seem appropriate to rely on the ordinary assumptions of independence and homoscedasticity. It is therefore not justified to derive the variance-covariance matrix of the ϑ-estimate from an estimate of the residual variance. We will use definition (9.13) or, rather, its finite form (9.15) which allows for multivariate non-normal models. The results are reported in Table 9 and it will be observed that the last coordinate, the acid concentration, appears to be more significant than the ordinary multivariate normal theory indicates (partial F-test score of 2.950 rather than 0.947); the exact level of significance cannot be inferred anymore because we have discarded the normality assumptions. The reported correlation matrix has been derived from the variance-covariance matrix in the ordinary way. The estimator of the error standard deviation remains the ordinary estimate.

The various quasi-robust regression fits we obtained have resulted in the sets of residuals recorded in Table 10. To allow comparison with the standard methods, we have included the ordinary least squares fits

Brownlee's stackloss Data
Least squares Results

Data set

i =	1	2	3	4	5	6	7	8	9	10	11	12	13	14	15	16	17	18	19	20	21
y	42	37	37	28	18	18	19	20	15	14	14	13	11	12	8	7	8	8	9	15	15
j = 1	1	1	1	1	1	1	1	1	1	1	1	1	1	1	1	1	1	1	1	1	1
j = 2	80	80	75	62	62	62	62	62	58	58	58	58	58	58	50	50	50	50	50	56	70
j = 3	27	27	25	24	22	23	24	24	23	18	18	17	18	19	18	18	19	19	20	20	20
j = 4	89	88	90	87	87	87	93	93	87	80	89	88	82	93	89	86	72	79	80	82	91

	j = 1	j = 2	j = 3	j = 4
Estimator	-39.920	.71564	1.2953	-.15212
Standard deviations "normal"	--	.13486	.36802	.15629
by (9.15)	6.5700	.16287	.45756	.08856
Partial F-tests	36.919	19.306	8.0140	2.9504

Correlation matrix

1	.20344	.11533	-.78704
.20344	1	-.87135	.02070
.11533	-.87135	1	-.25772
-.78704	.02070	-.25772	1

Error standard deviation : 3.2434

Table 9

on size $n = 21$ and on size $n = 17$, the 17 remaining observations when the reputed outliers 1, 3, 4 and 21 have been rejected; these two least squares sets of residuals are referred to as Fit 1 and Fit 34. The regressions have been obtained according to (9.24) and the ρ-functions have the definitions (6.24 - 31). Our specific definition (9.20) of the scale

Fit Residuals of Brownlee's Stackloss Data

Fit #	E Tab.6	1	2	3	4	5	6	7	8	9	10	11	12	13	14	15	16	17	18	19	20	21
Least powers (6.24)																						
1	1.00	3.2	-1.9	4.6	5.7	-1.7	-3.0	-2.4	-1.4	-3.1	1.3	2.6	2.8	-1.4	-0.1	2.4	0.9	-1.5	-0.5	-0.6	1.4	-7.2
2	0.95	3.6	-1.5	4.7	6.5	-1.5	-2.5	-1.7	-0.7	-2.4	0.7	1.9	1.8	-2.0	-0.5	2.2	0.8	-1.1	-0.1	0.0	1.6	-8.2
3	0.90	4.0	-1.1	4.9	6.9	-1.3	-2.2	-1.3	-0.3	-2.0	0.4	1.6	1.4	-2.3	-0.7	2.1	0.7	-1.0	-0.1	0.3	1.6	-8.5
4	0.85	4.4	-0.7	5.1	7.2	-1.3	-2.0	-1.1	-0.1	-1.7	0.2	1.3	0.9	-2.5	-1.0	1.9	0.5	-0.9	-0.1	0.3	1.6	-8.8
5	0.80	4.8	-0.3	5.3	7.4	-1.2	-1.9	-1.0	0.0	-1.6	0.1	0.9	0.5	-2.7	-1.3	1.6	0.3	-0.7	-0.1	0.4	1.6	-9.1
Function of Huber (6.25)																						
6	0.95	2.9	-2.2	4.1	6.2	-1.7	-2.8	-2.0	-1.0	-2.6	0.7	1.9	1.8	-2.1	-0.6	2.3	0.9	-1.0	-0.1	0.0	1.5	-8.6
7	0.90	3.1	-2.0	4.2	6.6	-1.5	-2.5	-1.7	-0.7	-2.2	0.5	1.6	1.4	-2.3	-0.8	2.3	0.9	-0.8	0.1	0.3	1.6	-9.0
8	0.85	3.9	-1.2	4.8	7.0	-1.4	-2.2	-1.4	-0.4	-1.9	0.4	1.4	1.1	-2.4	-1.0	2.0	0.7	-0.7	0.1	0.4	1.6	-8.9
9	0.80	4.4	-0.8	5.1	7.3	-1.2	-2.0	-1.0	0.0	-1.6	0.2	1.2	0.8	-2.6	-1.1	1.9	0.5	-0.7	0.0	0.4	1.6	-9.0
Modified Huber function (6.26)																						
10	0.95	3.1	-2.0	4.3	6.3	-1.7	-2.7	-1.9	-0.9	-2.5	0.7	1.8	1.7	-2.1	-0.6	2.3	0.9	-1.0	-0.1	0.0	1.5	-8.5
11	0.90	3.3	-1.8	4.4	6.7	-1.6	-2.5	-1.6	-0.6	-2.1	0.4	1.5	1.3	-2.3	-0.8	2.2	0.8	-0.8	0.1	0.3	1.6	-8.9
12	0.85	3.9	-1.2	4.8	7.0	-1.4	-2.2	-1.3	-0.3	-1.9	0.3	1.4	1.0	-2.4	-1.0	2.0	0.7	-0.7	0.1	0.4	1.6	-9.0
13	0.80	4.4	-0.7	5.1	7.3	-1.2	-2.0	-1.0	0.0	-1.6	0.2	1.2	0.8	-2.6	-1.1	1.8	0.5	-0.7	0.0	0.4	1.7	-9.0
Fair (6.27)																						
14	0.95	3.5	-1.7	4.6	6.4	-1.6	-2.6	-1.8	-0.8	-2.4	0.7	1.9	1.7	-2.0	-0.6	2.2	0.8	-1.0	-0.1	0.0	1.5	-8.3
15	0.90	3.8	-1.3	4.7	6.8	-1.4	-2.3	-1.4	-0.4	-2.0	0.4	1.5	1.3	-2.3	-0.8	2.1	0.7	-0.9	-0.1	0.2	1.6	-8.7
16	0.85	4.2	-0.9	5.0	7.1	-1.3	-2.1	-1.2	-0.2	-1.8	0.3	1.3	0.9	-2.5	-1.0	1.9	0.5	-0.8	0.0	0.3	1.6	-8.9
17	0.80	4.6	-0.5	5.3	7.4	-1.2	-1.9	-1.0	0.0	-1.6	0.2	1.1	0.6	-2.6	-1.2	1.7	0.4	-0.7	0.0	0.4	1.6	-9.0
Least squares on size n = 17																						
34	1.00	6.2	1.2	6.4	8.2	-0.7	-1.2	-0.4	0.6	-1.1	0.4	1.0	0.5	-2.5	-1.3	1.3	0.1	-0.4	0.1	0.6	1.9	-8.6

Table 10a

156

Fit Residuals of Brownlee's Stackloss Data

Fit #	E Tab.6	1	2	3	4	5	6	7	8	9	10	11	12	13	14	15	16	17	18	19	20	21
Function of Cauchy (6.28)																						
18	0.95	3.2	-2.0	4.3	6.3	-1.7	-2.7	-1.9	-0.9	-2.5	0.7	1.8	1.7	-2.1	-0.6	2.2	0.8	-1.0	-0.1	0.0	1.5	-8.5
19	0.90	3.3	-1.9	4.3	6.7	-1.6	-2.5	-1.6	-0.6	-2.1	0.4	1.5	1.2	-2.4	-0.9	2.2	0.8	-0.8	0.1	0.3	1.6	-9.1
20	0.85	3.5	-1.6	4.4	7.0	-1.5	-2.3	-1.3	-0.3	-1.8	0.2	1.2	0.9	-2.6	-1.1	2.1	0.7	-0.7	0.1	0.5	1.6	-9.3
21	0.80	3.9	-1.2	4.7	7.2	-1.4	-2.1	-1.1	-0.1	-1.6	0.1	1.1	0.7	-2.7	-1.2	2.0	0.6	-0.6	0.2	0.6	1.6	-9.4
Function of Welsch (6.29)																						
22	0.95	3.1	-2.1	4.2	6.3	-1.7	-2.7	-2.0	-1.0	-2.5	0.7	1.9	1.7	-2.1	-0.6	2.3	0.9	-1.0	-0.1	0.0	1.5	-8.5
23	0.90	2.9	-2.2	4.0	6.6	-1.7	-2.5	-1.6	-0.6	-2.1	0.2	1.3	1.0	-2.5	-1.0	2.2	0.8	-0.7	0.2	0.4	1.5	-9.4
24	0.85	2.8	-2.3	3.9	6.9	-1.7	-2.4	-1.4	-0.4	-1.8	-0.1	1.0	0.6	-2.8	-1.3	2.2	0.8	-0.5	0.3	0.7	1.6	-10.0
25	0.80	2.9	-2.2	3.9	7.1	-1.6	-2.2	-1.2	-0.2	-1.6	-0.2	0.8	0.3	-3.0	-1.4	2.2	0.8	-0.4	0.4	0.9	1.6	-10.2
Bisquare (6.30)																						
26	0.95	3.0	-2.1	4.2	6.2	-1.7	-2.8	-2.0	-1.0	-2.6	0.7	1.9	1.8	-2.1	-0.6	2.3	0.9	-1.0	-0.1	0.0	1.5	-8.5
27	0.90	2.7	-2.5	3.8	6.6	-1.8	-2.6	-1.7	-0.7	-2.1	0.1	1.2	0.9	-2.7	-1.2	2.2	0.8	-0.6	0.2	0.5	1.5	-9.7
28	0.85	2.5	-2.6	3.6	7.0	-1.8	-2.4	-1.4	-0.4	-1.7	-0.3	0.8	0.3	-3.0	-1.4	2.2	0.9	-0.4	0.4	0.9	1.5	-10.5
29	0.80	2.7	-2.4	3.7	7.2	-1.6	-2.2	-1.1	-0.1	-1.5	-0.3	0.7	0.2	-3.1	-1.4	2.2	0.9	-0.3	0.5	1.0	1.6	-10.5
Function of Andrews (6.31)																						
30	0.95	3.0	-2.1	4.2	6.2	-1.7	-2.8	-2.0	-1.0	-2.6	0.7	1.9	1.8	-2.1	-0.6	2.3	0.9	-1.0	-0.1	0.0	1.5	-8.5
31	0.90	2.6	-2.5	3.7	6.6	-1.8	-2.6	-1.7	-0.7	-2.1	0.1	1.2	0.8	-2.7	-1.2	2.2	0.8	-0.6	0.2	0.5	1.5	-9.8
32	0.85	2.5	-2.6	3.6	7.0	-1.8	-2.4	-1.4	-0.4	-1.7	-0.3	0.8	0.3	-3.0	-1.4	2.2	0.9	-0.4	0.4	0.9	1.5	-10.5
33	0.80	2.7	-2.4	3.7	7.2	-1.6	-2.2	-1.1	-0.1	-1.5	-0.3	0.7	0.2	-3.1	-1.4	2.2	0.9	-0.3	0.5	1.0	1.6	-10.5
Least squares on size n = 17																						
34	1.00	6.2	1.2	6.4	8.2	-0.7	-1.2	-0.4	0.6	-1.1	0.4	1.0	0.5	-2.5	-1.3	1.3	0.1	-0.4	0.1	0.6	1.9	-8.6

Table 10b

157

estimator makes difficult the comparison with the results derived by other robust estimation methods; as it is customary to use *MAD* with the methods of Huber and Andrews, we give in Table 11 the corresponding tuning constants. The information on the various regressions is condensed in Table 10 in such a compact form that it is hardly possible to skim it over to obtain a clear view. Let us first of all try to establish an overview and then consider a particular fit.

Tuning constants, the scale being estimated by MAD

Huber´s method

$$\begin{array}{lll} \text{Fit} & 6 & k = 2.2956 \\ & 7 & 1.9131 \\ & 8 & 1.6624 \\ & 9 & 1.3811 \end{array}$$

Andrews´s method

$$\begin{array}{lll} \text{Fit} & 30 & c = 2.35 \quad * \\ & 31 & 2.15 \quad * \\ & 32 & 1.9480 \\ & 33 & 1.80 \quad * \end{array}$$

* This value is approximate; it corresponds with two different Andrews´s fits which are interpolated by our own fit.

Table 11

The sets of residuals reported in Table 10 are derived by methods having similar powers and, therefore, are a good basis for comparing the methods. The efficiency values appearing in the second column are the same as those reported in Table 6, however in the present case they cannot be seen anymore as relative asymptotic efficiencies with respect to the normal distribution; the reason for this distinction lies in the more-than-one dimension feature. In order to compare the methods, it is convenient to examine how a given observation residual varies between the methods and with different efficiencies; this has been done for two observations in Fig. 7. The first and the fourth observations have been selected but a similar pattern exists for most of the observation pairs. The piecewise linear curves link together all the points representative of a given method. All the curves converge at an efficiency $E = 1$ (which is Fit 1) and terminate at an efficiency $E = .8$ (Fits 5, 9, ..., 33). Fig. 7 clearly indicates that the four quasi-robust methods produce similar results; they are, ranked according to their respective performances, the least powers, Fair (a very close second prize), modified Huber and Huber. Then there follow the four methods with descending influence functions, Cauchy (still a little robust), Welsh, Bisquare and finally Andrews' method which tends to maintain observation 2 at a certain distance from the regression hyperplane although it should not. Fig. 7 is helpful in comparing two columns of Table 10; Fig. 8 displays the relationship between two rows. Rather than comparing methods we now compare the observations with respect to their role in linear regression. We have selected rows 1 and 5, the two extreme least powers fits, but very similar displays correspond to most other pair selections. In fact, Fig. 8 is a curvilinear projection of the observations x_i in their 4-dimensional space onto a 2-dimensional subspace. Inspection reveals that observations 1, 3, 4 and 21 are outliers; moreover, in the 45-degree sectors where they lie, the residuals increase with increasing robustness and this confirms their outlier qualification. Observation 13 could also be regarded as critical but its residuals are too moderate to pretend at an outlier qualification.

Let us now turn our attention to a particular regression fit, namely Fit 17 obtained with the ρ-function Fair (6.27) at the nominal efficiency $E = .8$. The results have been derived using system (9.24) and the covariance matrix has been derived according to (9.8) and (9.23) neglecting the A_{12} factor. The essential data are reported in Table 12 which is to be compared with Table 9. Obviously the scatters of the $\hat{\vartheta}$-components are much smaller in the quasi-robust method than in the ordinary least squares method; this feature results from the limited influence of the observations that yield large residuals. However it should be noted that the estimate of the error standard deviation has

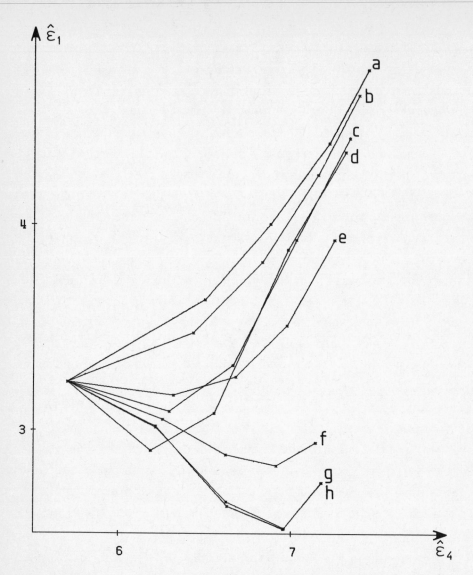

Fig.7. Brownlee's stackloss Data. Plot of $\hat{\varepsilon}_1$ against $\hat{\varepsilon}_4$ for 8 ρ-functions and at nominal efficiencies 1, 0.95, 0.9, 0.85 and 0.8. (a) Least powers (6.24); (b) Fair (6.27); (c) Modified Huber function (6.26); (d) Function of Huber (6.25); (e) Function of Cauchy (6.28); (f) Function of Welsch (6.29); (g) Bisquare (6.30); (h) Function of Andrews (6.31).

only been slightly reduced. This reduction is in a way an index of the outlier existence for it indicates whether the error terms have a distribution with tails that are more or less heavier than those for the normal law. The fact that the correlation matrices of Tables 9 and 12 differ

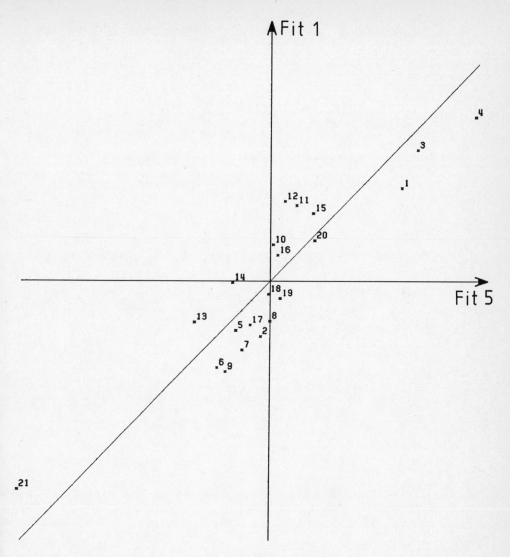

Fig.8. Brownlee's stackloss Data. Plot of the residuals of Fit 1 (Least squares method) against Fit 5 residuals (Least powers, $\nu = 1.22$).

markedly should be attributed to changes in the observation influence functions; these changes have been recognized but they will be more easily described in the next example.

Various disciplines are considering technical problems for which robust linear regression can be of some help although sometimes in an odd way. For image processing, Fischler and Bolles (1981) have

Data set

i =	1	2	3	4	5	6	7	8	9	10	11	12	13	14	15	16	17	18	19	20	21
y	42	37	37	28	18	18	19	20	15	14	14	13	11	12	8	7	8	8	9	15	15
j = 1	1	1	1	1	1	1	1	1	1	1	1	1	1	1	1	1	1	1	1	1	1
j = 2	80	80	75	62	62	62	62	62	58	58	58	58	58	58	50	50	50	50	50	56	70
j = 3	27	27	25	24	22	23	24	24	23	18	18	17	18	19	18	18	19	19	20	20	20
j = 4	89	88	90	87	87	87	93	93	87	80	89	88	82	93	89	86	72	79	80	82	91

	j = 1	j = 2	j = 3	j = 4
Estimator	-38.589	.82594	.69431	-.09961
Standard deviations	4.3975	.10416	.27537	.08105
Partial F-tests	77.005	62.876	6.3572	1.5105

Correlation matrix

1	-.45185	.54273	-.71659
-.45185	1	-.64296	-.02096
.54273	-.64296	1	-.62863
-.71659	-.02096	-.62863	1

Error standard deviation : 3.1908

Table 12

considered the problem of finding the best line passing through a set of points, most of them lying exactly on a straight line. It happened that the ordinary least squares regression had failed dramatically; then they tried out a sequence of linear least squares regressions combined with rejection at each step of the observation responsible for the largest residual. All these procedures failed and, finally, the authors recommended the application of a kind of bootstrap method and a majority rule. Their data set is reported in Table 13 from which it can be seen that the required solution is

$$y = z$$

or

$$\vartheta_1 = 0, \ \vartheta_2 = 1$$

which is a fit passing through 5 out of the 7 observations. Observation 4 is a little and observation 7 is far outlying. The ordinary least squares results of Table 13 reveal that observations 1 and 6 are responsible for the two largest residuals although they lie on the desired straight line; these two observations are the extreme points of the straight line and the fitted line makes an angle of 36° with it.

Naively it may be thought that robust regression can improve the situation. The results obtained with the ρ-function Fair at nominal efficiency $E = .8$ are reported in Table 14. It is seen that the residual of observation 7 has been reduced by a factor 2 although we could have expected a serious increase. The situation has grown worse. Possibly the trouble can be attributed to our robust method, so that we shall examine the results of other methods. These are summarized in Table 15 showing that all fits except for the last one miss the mark. Moreover they are consistently worse than the least squares solution. The last fit has been obtained with the Andrews' function by initializing the algorithm on the desired solution. This is a feature of all descending functions; they give you "the" good solution when you have been clever enough to provide them with "the" solution.

In fact we are faced here with an ill-posed problem. We have tried to obtain a fit by minimizing some metric, whereas the interest was in some majority rule. In order to see that the metric is not appropriate, consider the least squares solution that corresponds with the Euclidean metric; the sum of squared residuals with respect to the least squares fit is $SSE = 8.4234$ which is much better than with respect to the

Data set

i =	1	2	3	4	5	6	7
y	0	1	2	2	3	4	2
j = 1	1	1	1	1	1	1	1
j = 2	0	1	2	3	3	4	10

	j = 1	j = 2
Estimator	1.4820	.15766
Standard deviations	.64732	.15072

Correlation matrix

$$\begin{matrix} 1 & -.72327 \\ -.72327 & 1 \end{matrix}$$

Error standard deviation : 1.2979

Residuals		Influence function	
		j = 1	j = 2
i = 1	-1.4820	3.2477	-.53738
2	-.63964	1.1698	-.16135
3	.20270	-.29721	.02876
4	.04504	-.04971	.00142
5	1.0450	-1.1533	.03295
6	1.8874	-1.3985	-.14878
7	-1.0586	-1.5187	.78438

Table 13

desired solution ($SSE = 65$). Whatever you do, robust techniques never solve irrelevant problems.

To conclude this section on robust linear regression we would like to pay attention to a point which is often ignored. It might be believed

Fischler and Bolles's Data
Robust Results (Fair, E = 0.8)

Data set

```
    i =    1    2    3    4    5    6    7
    ------+--------------------------------
    y  |   0    1    2    2    3    4    2
       |
  j = 1 |   1    1    1    1    1    1    1
  j = 2 |   0    1    2    3    3    4   10
```

	j = 1	j = 2
Estimator	1.6855	.08140
Standard deviations	.88256	.24088

Correlation matrix

$$1 \qquad -.88867$$

$$-.88867 \qquad 1$$

Error standard deviation : 1.4697

		Influence function	
Residuals		j = 1	j = 2
i = 1	-1.6855	3.7195	-.77488
2	-.76691	2.5398	-.45937
3	.15169	-.92507	.12376
4	.07029	-.34374	.01182
5	1.0703	-1.2958	.04457
6	1.9889	-.63282	-.19835
7	-.49951	-3.0620	1.2524

Table 14

that the value of the influence function for most of the outlying observations is reduced by robust methods, but the actual situation is more
complex. Indeed compare the influence functions reported in Tables 13
and 14. The largest residual is due to observation 6; its two-component

Data set

i =	1	2	3	4	5	6	7
y	0	1	2	2	3	4	2
j = 1	1	1	1	1	1	1	1
j = 2	0	1	2	3	3	4	10

Method	E	j = 1	j = 2	Starting set
Least squares	1	1.4820	.15766	Any
Least powers	0.8	1.7918	.06884	Any
Huber	0.8	1.5157	.10600	Any
Modified Huber	0.8	1.5472	.10193	Any
Fair	0.8	1.6855	.08140	Any
Cauchy	0.8	1.5107	.11812	Least squares result
Welsch	0.8	1.4802	.13504	Least squares result
Bisquare	0.8	1.4766	.14085	Least squares result
Andrews	0.8	1.4771	.14099	Least squares result
Andrews	0.8	.00055	.92720	(0.0, 1.0)'

Table 15

influence function exhibits a fine decrease for the first component while the second increases, although to a moderate extent. Observation 1 has an increased influence function for both components and it has the second largest residual which has increased as well. These alterations might appear inconsistent when one only takes into account the last factor z_i of (9.25), because they are accounted by the first factor, the inverse matrix. Even for very outlying observations, the influence function cannot be very large when the sample size is moderate. This is demonstrated on the data set by changing y_7 into 10. Then six observations lie on a straight line and observation 4 becomes a very neat single outlier. The results reported in Tables 16 and 17 indicate that the influence function values have very much decreased with the robust procedure, but the magnitude of the outlier influence function has not change much with respect to the other influence function magnitudes in spite of a 16-fold increase in residual ratio increase. For instance from Tables 16 and 17 we have, with respect to the first component and the first observation,

$$[\,\Psi_{1,4}\,/\,\Psi_{1,1}\,]_{Fair}\,/\,[\,\Psi_{1,4}\,/\,\Psi_{1,1}\,]_{lsq}\,=\,1.0920$$

and

$$[\,\hat{\varepsilon}_4\,/\,\hat{\varepsilon}_1\,]_{Fair}\,/\,[\,\hat{\varepsilon}_4\,/\,\hat{\varepsilon}_1\,]_{lsq}\,=\,15.703.$$

9.6. Robust estimation of multivariate location and scatter

If we were asked to rank the problems we are faced with in robust estimation according to their respective difficulties, we would certainly place the multivariate location problem among the most difficult ones. Such a ranking could be the one-dimensional location-scatter problem followed by the linear regression group, then the multivariate location-scatter problem and finally all non-linear questions. The one-dimensional location-scatter problem is the object of most of the attention because in essence it already includes most of the difficulties; moreover it constitutes a framework where the asymptotic theories are more or less manageable. Its generalization to the linear regression problem presents difficulties which can be handled indeed, but the generalization from one dimension to several dimensions seems to render the location-scatter problem unmanageable. The difficulties are certainly as much of a conceptual as of a technical nature. We just are uncertain about the objectives. Most of us have serious difficulties in imagining three-dimensional spaces when faced with the plan drawings of a house; except for the most imaginative among us, we do not know how to represent a data structure occurring in 4, 5 or more dimensions. The only tool we have is the projection; we may project points onto a plane but their relationships (distances, angles, volumes) become very obscure, and fade out with increasing dimensionalities.

Modified Fischler and Bolles´s Data
Least squares Results

Data set

i =	1	2	3	4	5	6	7
y	0	1	2	2	3	4	10
j = 1	1	1	1	1	1	1	1
j = 2	0	1	2	3	3	4	10

	j = 1	j = 2
Estimator	-.15766	1.0045
Standard deviations	.16900	.01762

Correlation matrix

$$1 \qquad -.58950$$

$$-.58950 \qquad 1$$

Error standard deviation : .41373

Residuals		Influence function	
		j = 1	j = 2
i = 1	.15766	-.34550	.05717
2	15315	-.28009	.03863
3	.14865	-.21795	.02109
4	-.85586	.94453	-.02699
5	.14414	-.15908	.00454
6	.13964	-.10347	-.01101
7	.11261	.16156	-.08344

Table 16

With regard to the definition of a location estimator however, the situation is rather pleasant. We define a location estimator as "the" location such that its "distances" to all observations are minimized - Each word of this definition could lead to endless discussions - How to

Modified Fischler and Bolles's Data
Robust Results (Fair, E = 0.8)

Data set

i =	1	2	3	4	5	6	7
y	0	1	2	2	3	4	10
j = 1	1	1	1	1	1	1	1
j = 2	0	1	2	3	3	4	10

	j = 1	j = 2
Estimator	-.01161	1.0004
Standard deviations	.01722	.00153

Correlation matrix

$$1 \qquad -.65948$$

$$-.65948 \qquad 1$$

Error standard deviation : .31280

Residuals		Influence function j = 1	j = 2
i = 1	.01161	-.03285	.00494
2	.01123	-.02715	.00342
3	.01085	-.02172	.00197
4	-.98953	.09807	-.00357
5	.01047	-.01654	.00060
6	.01009	-.01164	-.00069
7	.00782	.01184	-.00668

Table 17

define a multivariate scatter is not only vague, but even unknown.
There have been quite a few proposals, some of them being serious con-
tenders, others being just rubbish. They can also be heuristic and
based on common sense. Huber (1977a) bases his approach on a

generalization of the concentration ellipse, Rey (1979) recommends to derive a scatter estimate from the variance-covariance matrix of the location estimator but, in both cases, the consistency between the definitions for location and scatter estimations seems rather artificial. Let us try to be a little logical and pay attention to Fisher's views on the subject (1922) :

> In order to arrive at a distinct formulation of statistical problems, it is necessary to define the task which the statistician sets himself: briefly, and in its most concrete form, the object of statistical methods is the reduction of data. A quantity of data, which usually by its mere bulk is incapable of entering the mind, is to be replaced by relatively few quantities which shall adequately represent the whole, or which, in other words, shall contain as much as possible, ideally the whole, of the relevant information contained in the original data.... Since the number of independent facts supplied in the data is usually far greater than the number of facts sought, much of the information supplied by any actual sample is irrelevant. It is the object of the statistical processes employed in the reduction of data to exclude this irrelevant information, and to isolate the whole of the relevant information contained in the data.

Let us start with a definition of the location estimator and, then, we will gradually succeed in defining a corresponding multivariate scatter estimate. The final set of equation strongly reminds us of the fundamental paper of Maronna (1976) and all his asymptotic properties remain valid; they are further analyzed by Carroll (1978).

The location estimator minimizes distances. Very naturally we may define these distances by the Mahalanobis metric, say

$$u_i = [\, (\, x_i - \vartheta\,)'S^{-1}(\, x_i - \vartheta\,)\,]^{1/2} \tag{9.26}$$

where s is the multivariate matrix of variance-covariance; in doing so we may assume that s is known a priori. Minimizing the distances is equivalent to minimizing any continuously increasing transform of these distances. For convenience and to remain consistent with the multivariate normal theory, we will define $\hat{\vartheta}$ as a quantity for which the sum of the transformed distances is a minimum, i. e.

$$M = \sum \rho\,(u_i) = \min \text{ for } \vartheta. \tag{9.27}$$

Let us now consider the scatter estimation. We have a few prior ideas which can be seen as boundary conditions in defining the estimator. It should concur with the ordinary definition whenever the ρ-function of (9.27) is the square function and for the one-dimensional situation we must have

$$1 = \frac{1}{(n-1)\, Int} \sum \rho\, (u_i)$$

according to (9.24) with $g = 1$. The role of the ρ-function being to down-weight the large distances, we select a weighted form for the s-matrix and, as does Campbell (1980), then identify the weights. Let the matrix be given by

$$S = \frac{1}{(n-1)\, Int} \sum w_i\, (x_i - \vartheta)\, (x_i - \vartheta)';$$

then, in the least squares situation, it should equate

$$\frac{1}{n-1} \sum (x_i - \vartheta)\, (x_i - \vartheta)'$$

with $Int = 1$ by (9.21), hence $w_i = 1$; in the one dimensional case we obtain

$$w_i = \rho\, (u_i)\, /\, (x_i - \vartheta)^2.$$

These two boundary conditions are satisfied if

$$w_i = \rho\, (u_i)\, /\, u_i^2$$

and this leads us to define the multivariate scatter matrix by

$$S = \frac{1}{(n-1)\, Int} \sum [\, \rho\, (u_i)\, /\, u_i^2\,]\, (x_i - \vartheta)\, (x_i - \vartheta)'. \tag{9.28}$$

Therefore the solution of the multivariate location scatter problem is given by the MM estimators (9.26 - 28).

The definition of the multivariate scatter has been handled in a way which conceals part of the questions. Although the large sample properties are satisfactory, it may be difficult to understand the finite sample behavior anyway; for instance, the sensitivity of a given observation on the principal component orientation may be unexpected although it can be understood a posteriori. Obviously experience is still lacking in this domain. Another aspect which constrains the possible interest of simultaneous robust estimations of location and scatter is the near-impossibility of obtaining a variance-covariance matrix for the location estimator. Finally the most easily conceived algorithms are of the relaxation type and have a strong tendency to "oscillate" (see § 9.8). Their damping requires special techniques which are essentially heuristic.

The situation in the domain of multivariate location-scatter robust estimation is so prehistoric that we hesitate to propose here any example for further consideration.

9.7. Robust non-linear regression

To a novice, it seems that non-linear and linear regressions are not very different. And, indeed, they do not differ very much, although the little non-linear regression has not in common with its notorious linear brother is sufficient to ruin all theoretical works from time to time and to limit seriously the applications. The multivariate location-scatter problem could be involved; this problem is hazardous.

The similarity between the linear and non-linear approaches is so important that it permits a very concise presentation. The non-linear model corresponding to (9.9) can be written as

$$y_i = f(z_i, \vartheta) + \varepsilon_i \tag{9.29}$$

where ϑ is a g-component vector, and an observation remains defined by the set (9.10), i.e.

$$x_i = (y_i, z'_i)'.$$

As before we need not decide whether the error term ε_i is due to the y_i or to the z_i component. The previous discussion on robustness, independency and scale estimation still holds good and therefore we may define robust estimators of non-linear regression by (9.24) or rather by

$$s^2 \sum \rho \{[\, y_i - f\,(z_i, \vartheta)\,]\, /\, s\} = \min \text{ for } \vartheta,$$

$$1 = \frac{1}{(n-g)\ Int} \sum \rho \{[y_i - f\,(z_i, \vartheta)\,]\, /\, s\}. \tag{9.30}$$

Thus far we have not restricted the family of function $f\,(., .)$ we have been considering. The fact is that it seem impossible to specify this family; we will require the functions to behave well but nobody knows exactly what is meant by such a statement. However a few properties of regularity and smoothness in the vicinity of the estimated $\hat{\vartheta}$ are welcome; say we expect $f\,(., \vartheta)$ to be continuous (or piecewise continuous) and to be of bounded variation in the vicinity of $\hat{\vartheta}$. The contrary is very subjective as well; a function is ill-behaving whenever the theory is not applicable, the estimating algorithms fail, or whenever the statistician feels challenged. Along these lines there may be a lot of philosophy to be developed but we would prefer to remain a little more pragmatic and refer the reader to textbooks on non-linear estimation (e. g. Bard, 1974). We assume that the second derivatives

$$\frac{\partial^2}{\partial \vartheta^2} f\,(z_i, \vartheta) \approx 0 \tag{9.31}$$

in the sense that the terms where they appear can be neglected.

We now evaluate the influence functions according to (9.23) where, as we recall, the first equation of (9.30) is

$$M_1 = \sum \rho_1\,(x_i, \vartheta_1, \vartheta_2) = \min \text{ for } \vartheta_1 .$$

ϑ_1 stands for ϑ and ϑ_2 for s^2, the squared scale. If we abbreviate the notation, we have

$$u = [y - f\,(z, \vartheta_1)]\, \vartheta_2^{-1/2}$$

$$(\partial / \partial \vartheta_1)\, u = [-(\partial / \partial \vartheta_1)\, f\,(z, \vartheta_1)]\, \vartheta_2^{-1/2}$$

$$= -\vartheta_2^{-1/2} \dot{f} .$$

$$(\partial / \partial \vartheta_2) \, u = [y - f(z, \vartheta_1)] \left[-\frac{1}{2} \vartheta_2^{-3/2} \right]$$

$$= -\frac{1}{2} u / \vartheta_2 \ ,$$

$$\dot{\rho}_1(x, \vartheta_1, \vartheta_2) = (\partial / \partial \vartheta_1) \, \rho_1(x, \vartheta_1, \vartheta_2)$$

$$= (\partial / \partial \vartheta_1) \, [\vartheta_2 \, \rho(u)]$$

$$= \vartheta_2 \, [(d / du) \, \rho(u)] \, [(\partial / \partial \vartheta_1) u]$$

$$= -\vartheta_2^{1/2} \, \dot{\rho}(u) \, \dot{f} \ ,$$

$$(\partial / \partial \vartheta_1) \, \dot{\rho}_1(x, \vartheta_1, \vartheta_2) = \ddot{\rho}(u) \dot{f} \, \dot{f} \ ,$$

$$(\partial / \partial \vartheta_2) \, \dot{\rho}_1(x, \vartheta_1, \vartheta_2) = \frac{1}{2} \vartheta_2^{-1/2} \, [u \, \ddot{\rho}(u) - \dot{\rho}(u)] \, \dot{f} \ ,$$

$$\dot{\rho}_2(x, \vartheta_1, \vartheta_2) = \vartheta_2 - K \, \vartheta_2 \rho(u)$$

$$(\partial / \partial \vartheta_2) \, \dot{\rho}_2(x, \vartheta_1, \vartheta_2) = 1 - K[\rho(u) - \frac{1}{2} u \, \dot{\rho}(u)] \ ;$$

and, with regard to the terms included in the summations,

$$\sum \dot{\rho}_1 (x, \vartheta_1, \vartheta_2) = -\vartheta_2^{1/2} \sum \dot{\rho}(u_i) \, \dot{f} = 0 \ ,$$

$$A_{11} = \frac{1}{n} \sum \ddot{\rho} (u_i) \, \dot{f} \dot{f}' \ ,$$

$$A_{12} = \frac{1}{2n} \vartheta_2^{1/2} \sum [u_i \, \ddot{\rho}(u_i) \, \dot{f}] \ ,$$

$$\sum \dot{\rho}_2 (x, \vartheta_1, \vartheta_2) = \vartheta_2 \sum [1 - K\rho(u_i)] = 0 \ ,$$

$$A_{22} = \frac{1}{2n} K \sum u_i \, \dot{\rho}(u_i) \ .$$

Substituting these expressions into (9.23) we obtain the influence function with respect to the variance estimator $s^2 = \vartheta_2$

$$\Psi_2(x_0) = 2ns^2 [\sum u_i \, \dot{\rho}(u_i)]^{-1} [\rho(u_0) - 1/K] \ . \tag{9.32}$$

as well as with respect to the estimated parameter set $\hat{\vartheta} = \vartheta_1$

$$\Psi_1(x_0) = ns \, [\sum \ddot\rho(u_i) \, \dot f \dot f']^{-1} \{\dot\rho(u_0) \, \dot f \, +$$

$$[1/K - \rho(u_0)] \, [\sum u_i \, \dot\rho(u_i)]^{-1} \, [\sum u_i \, \ddot\rho(u_i) \, \dot f]\} \, . \tag{9.33}$$

The last term between braces of (9.33) tends to cancel whenever the residuals u_i are symmetrically distributed. We will omit this term in order to obtain a compact form for the estimator variance-covariance matrix. According to (9.8) we obtain

$$Cov\,(\hat\vartheta) = \frac{ns^2}{n-1}[\sum \ddot\rho(u_i) \, \dot f \dot f']^{-1} \, [\sum \dot\rho(u_i)^2 \, \dot f \dot f'] \, [\sum \ddot\rho(u_i) \, \dot f \dot f']^{-1} \, . \tag{9.34}$$

which is an estimate presenting a relative error of order inferior to n^{-1}. It will be noted that the expression (9.34) does not permit separation of the terms related to the relative intrinsic curvature and those refer-ring to the parameter-effects curvature, as do Beale (1960), Bates and Watts (1980) and Clarke (1980). The main complication may be attri-buted to the imbedding of the non-linear model (9.29) into the non-linear robust minimization rule.

The general theory of non-linear estimation is progressing very slowly; miscellaneous practical items have been proposed in most cases to satisfy needs in econometry (see Bard, 1974) and one frequently resorts to asymptotic derivations (Ramsey, 1978) in order to get an insight into the behavior of the estimates. On finite samples, very little is known about the most non-linear models; however they are widely applied. In such a context the jackknife and bootstrap methods appear to be the only tools available to safely assess the estimated parameter set; unfortunately these methods are either horribly expensive or require a lot of tailoring. Robust methods are rarely considered on account of the difficulties already encountered with the least squares minimization criterion; a notable exception is the work of Bierens (1980, 1981) who has given a survey of and a structure to all the relevant theory.

Very few applications have been published in the domain of robust non-linear regression. Wahrendorf (1978) fits a curve $(g=4)$ of a hyperbolic type (its equation has not been reported correctly) and demonstrates with the help of plots and tables what degree of robustness has been achieved; the sample size of his last example is so low ($n=8$) that he has lost all the possibilities of assessment that are based on theory. We conclude this section by an application demonstrating the benefit of the robustness approach; the estimator efficiency has been increased by a factor of 9.

A set of twenty four similar resistors has been subjected to an aging test to assess how their values drift with time. These resistors have been specially manufactured so as to be very stable and therefore an appropriate testing equipment had to be designed; the data set which is reported in Table 18 relates to a part of the preliminary study which was aimed at assessing the capabilities of the testing equipment as well. Each row of Table 18 gives the measured values of a given resistor i after respectively 0, 120, 264, 576 and 1008 hours at a controlled temperature of 125° C. We denote by R_{it} the resistance of the i-th resistor. There are reasons to believe that this resistance does not drift anymore after some period of time, i. e.

$$R_{i\infty} = R_{i0}(1+\alpha)$$

where α is the relative drift. It may also be supposed that initially the drift changes rapidly and then slowly. A bilinear model seems appropriate

$$R_{it} = R_{i0}\left[\,1+\alpha\beta t/(\alpha+\beta t)\,\right]$$

where β is the drift rate at 0 hour. Thus, we have to estimate α and β as well as the 24 proportionality constants ($g=26, n=120$). It will be observed that our parametrization attempts to procure a great independency between the estimators as recommended by Ross (1970, 1978, 1980) in order to eventually obtain more or less elliptical confidence regions. The least squares results reported in Table 19 have been the most expensive ones as regards computer time because we did not succeed in achieving a rapid convergence to the solution; later on, we have understood that the poor convergence was due to the large coefficient

176

of variation of $\hat{\beta}$. It will be seen from Table 19 that the error standard deviation estimate is exceptionally high as compared to the variations appearing in the observation set; this estimate has been seriously inflated by a few outliers as well as by a moderate adequacy of the model. A quasi-robust fit has been very rapidly derived with the function Fair (6.27) at a nominal efficiency $E = 0.95$. The reduced sensitivity to large residuals appear to be very beneficial. The few items which have been singled out for Table 19 are typical of the information given by the (26×26)-variance-covariance matrix (9.34); they could be further analyzed.

Non-linear Estimation
Resistor Values

i	Time (hours)				
	0	120	264	576	1008
1	35184	35185	35186	35187	35189
2	39341	39374	39347	39349	39351
3	27496	27498	27498	27498	27500
4	27867	27869	27870	27870	27872
5	29710	29712	29713	29714	29715
6	30493	30494	30494	30495	30496
7	38162	38166	38167	38168	38170
8	36507	36513	36515	36515	36516
9	31132	31137	31139	31140	31141
10	30862	30869	30871	30871	30873
11	26375	26378	26440	26380	26381
12	29170	29172	29172	29173	29174
13	27793	27795	27798	27799	27800
14	28512	28520	28522	28523	28524
15	31474	31479	31370	31482	31483
16	34659	34667	34669	34669	34672
17	30566	30570	30572	30573	30574
18	30342	30347	30349	30351	30352
19	29462	29467	29580	29470	29472
20	30222	30230	30232	30232	30234
21	31577	31581	31584	31584	31586
22	28410	28417	28419	28420	28422
23	28005	28010	28012	28013	28014
24	28817	28823	28825	28826	28827

Table 18

Estimated values	Least squares	Fair, E = 0.95
α	0.000266	0.000273
$\sigma(\alpha)$	0.000070	0.000023
β	$6.53 \ 10^{-6}$	$3.46 \ 10^{-6}$
$\sigma(\beta)$	$10.8 \ 10^{-6}$	$1.67 \ 10^{-6}$
Error stand. dev	15.757	10.294
$\rho_{\alpha,\beta}$	-0.04858	-0.12145
$R_{1,0}$	35179.63	35179.96
$\sigma(R_{1,0})$	2.26	1.25
$R_{15,0}$	31451.73	31470.23
$\sigma(R_{15,0})$	19.37	5.68
$R_{24,0}$	28818.22	28818.60
$\sigma(R_{24,0})$	1.64	0.71

Table 19

9.8. Numerical methods

Quite a few numerical methods have been devised in order to be able to evaluate *MM* estimators. Evidently these methods are also applicable to the simpler structure of the *M* estimators. They will be presented in general terms with, however, a view to estimating the non-linear robust regression problem as defined by (9.30). Subcases can be obtained by making the appropriate substitutions; they are the linear regression and the one-dimensional location-scatter problems. The last-mentioned methods are widely applicable and are not as sensitive to the quality of the starting parameter set as are most of the other methods

We will not discuss all the problems involved in the initialization of the iterative algorithms because these problems are highly context dependent; they present any degree of difficulty, i. e. from trivial to near-unsolvable. Some ideas on the possible complexity can be derived from the review of Nirenberg (1981) on non-linear topologies. When a proper initial parameter set is available, a one-step Newton iteration may suffice as indicated by Bickel (1975) as well as by Hill and Holland (1977). From time to time specially developed algorithms can exhibit an outstanding performance as in the case of Huber's function (6.25) - see Huber (1973, § 8), Huber and Dutter (1974), Dutter (1977), Huber (1978) and Dutter and Huber (1981) - With regard to non-linear regression, it is often convenient to initialize the robust estimation procedure with the aid of a least squares procedure; the corresponding difficulties are treated by Gill and Murray (1978) as well as Ralston and Jennrich (1978). The survey of the most common software packages by Hiebert (1981) indicates that further investigation is badly needed in this domain.

9.8.1. Relaxation methods

Although many variations on these methods are possible, we will only present the main algorithm in which no damping factor appears. This will be sufficient to indicate where lie the difficulties and what should be taken into account in order to solve these difficulties.

The algorithm is iterative and consists in a treatment which is repeated one or more times until the convergence is deemed sufficient. Each repetition, or step, will be identified by an index k and we will assume that we have at our disposal ϑ_1^k and ϑ_2^k, approximations of ϑ_1 and

ϑ_2, when the step begins. Initially, we can enter the solution corresponding to the least squares method.

In order o solve the system (9.30)

$$s^2 \sum \rho \{[\ y_i - f(z_i, \vartheta)]/\ s\} = \min \text{ for } \vartheta$$

$$1 = \frac{1}{(n-g)\ Int} \sum \rho \{[\ y_i - f(z_i, \vartheta)]/\ s\}\ ,$$

we assume a prior knowledge of a set (ϑ^k, s_k); then with the help of one equation we improve one of the set components and, then, we introduce the corrected component into the other equation to complete the set, thus updating it into $(\vartheta^{k+1}, s^{k+1})$. In the present case, this scheme implies minor modifications of the equations. We start by modifying the scale estimate with the help of the second equation written in analogy to (9.20), as

$$(s^{k+1})^2 = (s^k)^2\ \frac{1}{(n-g)\ Int} \sum \rho \{[\ y_i - f(z_i, \vartheta^k)]/\ s^k\}\ .$$

We thus obtain an improved set (ϑ^k, s^{k+1}) which will now be inserted into the first equation. It will be observed that ϑ is now regarded as an M estimator because s^{k+1} is "given". After differentiating the first equation into

$$\sum \dot{\rho} \{[\ y_i - f(z_i, \vartheta)\ ]/s\}\ \dot{f}(z_i, \vartheta) = 0\ .$$

we have essentially two courses of actions to evaluate an improved ϑ^{k+1}: either we apply the Newton procedure (6.10) or we transform the equation into a structure of weighted least squares linear regression. We now describe this second possibility and will meet again with the Newton method in the next section. Denoting the normalized residual by u_i and its approximate value by u_i^*, we have

$$u_i = [\ y_i - f(z_i, \vartheta)\ ]/s^{k+1}\ ,$$

$$u_i^* = [\ y_i - f(z_i, \vartheta^k)\ ]/s^{k+1}\ ,$$

and the implicit equation

$$\sum \dot{\rho}(u_i)\, \dot{f}(z_i,\vartheta) = 0 \ .$$

Then we multiply and divide the terms by u_i, replace approximate values by exact ones and finally obtain a form typical of weighted least squares linear regression. We perform these steps as follows

$$\sum \dot{\rho}(u_i)\, \frac{u_i}{u_i}\, \dot{f}(z_i,\vartheta) = 0 \ ,$$

$$\sum \{\frac{\dot{\rho}(u_i)}{u_i}\}\, u_i\, \dot{f}(z_i,\vartheta) = 0 \ ,$$

$$\sum \{\frac{\dot{\rho}(u_i^*)}{u_i^*}\, [\, u_i^* + (u_i - u_i^*)\,]\, \dot{f}(z_i,\vartheta^k) \approx 0 \ ,$$

$$\sum w_i \{[\, y_i - f(z_i,\vartheta^k)\,] + [\,(\vartheta^k - \vartheta)'\, \dot{f}(z_i,\vartheta^k)\,]\}\, \dot{f}(z_i,\vartheta^k) \approx 0 \ ,$$

$$\{\sum w_i\, [\, \dot{f}(z_i,\vartheta^k)\,]\, [\, \dot{f}(z_i,\vartheta^k)\,]'\}\, (\vartheta^{k+1} - \vartheta^k)$$

$$= \sum w_i\, [\, y_i - f(z_i,\vartheta^k)\,]\, \dot{f}(z_i,\vartheta^k) \ .$$

The last equation is linear in $(\vartheta^{k+1} - \vartheta^k)$ and constitutes the definition of the improved estimate ϑ^{k+1}. Inasmuch as the linearization and the sensitivity of the weight-values

$$w_i = \dot{\rho}(u_i^*)/u_i^*$$

are not too critical, the new estimate ϑ^{k+1} may be rather good. This method has been recommended by Beaton and Tukey (1974); with regard to the implementation details, the algorithms have been studied by Wampler (1979), Coleman et al. (1980) and Birch (1980).

A significant drawback of this relaxation method is that it is not clear whether it converges or not. It may be observed that sometimes the starting set has the utmost importance for the final solution. Investigation of a few pathological situations has made it clear to the author that this can be attributed to the existence of several minima (and possibly saddle-points) in the joint parameter space $(theta_1, \vartheta_2) = (theta, s)$. Another weak aspect of the relaxation method is its low convergence speed. The splitting of the iteration in two successive steps limits the convergence to a linear one; this can sometimes be obviated by the implementation of a predictor-corrector method. However, in view of the risks involved, we prefer to implement an accelerating procedure only for scale estimation.

In spite of the obvious deficiencies of the above approach and frequently without having knowledge of the possible hazards of convergence, most experimenters have adopted a relaxation method. This is understandable when it is born in mind that most computer centers have at their disposal an efficient software to solve generalized least squares regression problems. The algorithms are generally denotes as "reweighted" least squares.

In most cases the users of reweighted least squares algorithms rely on the variance-covariance matrix which is finally issued by the package. This reliance is very dangerous because the weights in robust estimation are data-dependent and, consequently, cannot be regarded as constants known a priori. In fact the relaxation method is hardly favorable.

9.8.2. Simultaneous solutions

Various methods have been proposed to estimate simultaneously the two sets of parameters ϑ_1 and ϑ_2, in our specific case, ϑ and s. Here we turn our attention to Newton-Raphson iterations. The argument is quite general and can be presented for any number of equations; it is in many respects similar to the derivation of §9.2 leading to an expression of the influence function.

In all our evaluations of the approximate set (ϑ^k, s^k), we write the main equations in a linearized form. With

$$\delta\vartheta = \vartheta^{k+1} - \vartheta^k ,$$

$$\delta s = s^{k+1} - s^k ,$$

and omitting the second-order derivative of $f(z,\vartheta)$, we arrive at the abbreviated notation

$$\sum [\, \dot\rho(u_i)\, \dot f - \tfrac{1}{s}\ddot\rho(u_i)\, \dot f \dot f\,' \,\delta\vartheta - \tfrac{1}{s}\ddot\rho(u_i)\, u_i\, \dot f\, \delta s\,] = 0$$

and

$$\sum [\, \rho(u_i) - \tfrac{1}{s}\dot\rho(u_i)\, \dot f\,' \,\delta\vartheta - \tfrac{1}{s}\dot\rho(u_i)\, u_i\, \delta s\,] = (n-g)Int = K_1 \,.$$

It is convenient to write this system of two linear equations in the form

$$A\,\delta\vartheta + b\,\delta s = e$$

$$c'\,\delta\vartheta + d\,\delta s = g \tag{9.35}$$

with

$$A = \tfrac{1}{s}\sum \ddot\rho(u_i)\, \dot f \dot f\,' \,,$$

$$b = \tfrac{1}{s}\sum \ddot\rho(u_i)\, u_i\, \dot f \,,$$

$$c = \tfrac{1}{s}\sum \dot\rho(u_i)\, \dot f \,,$$

$$d = \tfrac{1}{s}\sum \dot\rho(u_i)\, u_i \,,$$

$$e = s\,c \,,$$

$$g = [\,\sum \rho(u_i)\,] - K_1 \,.$$

The fact that the second equation is scalar leads by gaussian elimination to

$$\delta\vartheta = (d\,A - b\,c')^{-1}(d\,e - g\,b)$$

and

$$ds = \tfrac{1}{d}(g - c'\,\delta\vartheta) \,;$$

without great profit we might write the matrix inverse with the help of (2.45), but a more efficient procedure consists in directly solving the system (9.35) rather than in performing a matrix inversion.

Thus the iteration step

$$\vartheta^{k+1} = \vartheta^k + \delta\vartheta ,$$

$$s^{k+1} = s^k + \delta s \tag{9.36}$$

results from a linearization of the involved non-linear functions. Whenever these functions have a sufficient regularity, the final convergence is quadratic in speed whenever the iterations do converge.

9.8.3. Solution of fixed-point and non-linear equations

In the preceding sections, we have seen that the relatively obvious methods of relaxation or of simultaneous solution used for the computation of MM estimators can be, and frequently are, rather time-consuming. We have also observed that the one-step method might be inaccurate and, possibly, could dissimulate divergence of the otherwise-iterative process. Even the use of M estimators can present difficulties (see § 6.4.2). This section deals with fixed-point considerations and the application of the continuation method to solve non-linear equations.

Basically all converging iterative methods can be seen as fixed-point computations. In our context of MM estimation, given an approximate solution $(\vartheta_1^\bullet,...,\vartheta_g^\bullet)$ we work out, using some arithmetic rule $R'(.)$, an "improved" approximate solution

$$(\vartheta_1^{\bullet\bullet}, \cdots ,\vartheta_g^{\bullet\bullet}) = R'(\vartheta_1^\bullet, \cdots ,\vartheta_g^\bullet) , \tag{9.37}$$

and repeat the process until stationarity is obtained, that is to say until

$$(\vartheta_1^\bullet, \cdots ,\vartheta_g^\bullet) = R'(\vartheta_1^\bullet, \cdots ,\vartheta_g^\bullet) .$$

Then, the solution is

$$(\vartheta_1, \cdots ,\vartheta_g) = R'(\vartheta_1^\bullet, \cdots ,\vartheta_g^\bullet) .$$

The solution of the general fixed-point problem still presents many difficulties. We will first consider the situation in which we are sure that

184

there is a single solution; then we shall be considering multiple solutions. Possibly the greatest advantage of the fixed-point approach is that it provides a closer insight into the multiplicity aspects as well as the capabilities of the continuation method. The generality and the newly-recognized importance of the fixed-point concept are well illustrated by the series of papers edited by Swaminathan (1976). Karamardian (1977), Balinski and Cottle (1978), Forster (1980) and Wold (1981). These cover mathematical theories and algorithmic features in various fields; it must be observed that there is a large body of literature relating to problems arising in econometry, where the appropriate techniques are similar to the Gauss pivoting method for solving systems of linear equations.

In order to solve the fixed-point problem we may simply iterate on the rule (9.37), but whether such a process will converge is not evident. We note that it is equally difficult to apply the Brouwer theorem as to define some contracting mapping through a Lipschitz constant, owing to the impossibility of demarcating a compact subset in the space of the parameters, except in fairly trivial situations - See Henrici (1974, § 6.12) for this subject - Thus, we will rather devote our attention to a general computation algorithm investigated by Scarf (1973) and further refined in Kellog et al. (1976). A few complementary aspects and practical considerations are reported by Todd (1976) as well as in the literature we mentioned.

This algorithm is based on the "continuation method" and consists in following "the" solution when some parameter varies. Assume we want the single solution of

$$\vartheta = R(\vartheta)$$

with

$$\vartheta = (\vartheta'_1, \cdots, \vartheta'_g) '$$

and assume that we know the solution for another rule $R_0(.)$

$$\vartheta_0 = R_0(\vartheta_0) ,$$

then ϑ is the solution for $\lambda = 1$ of

$$\vartheta = \lambda R(\vartheta) + (1-\lambda) R_0(\vartheta) . \tag{9.38}$$

The method consists in following the solution from $\lambda = 0$, where it is $\vartheta = \vartheta_0$, up to $\lambda = 1$ where it is the fixed point solution desired. The efficiency of the method is the higher as the implicit function $\vartheta(\lambda)$ with respect to the "variable" λ is the smoother.

The continuation of the homotopy path from $\lambda = 0$ up to $\lambda = 1$ may be achieved in several different ways. This is discussed by Ortega and Rheinboldt (1970) as well as Garcia and Gould (1980) who more specially consider Newton-like methods. Let us first show that no great difficulty is to be expected as long as some matrix condition is fulfilled. We examine whether there is a non-zero radius of convergence for any given λ as follows. The existence of such a convergence domain implies that a Newton-Raphson iteration has a finite step. Let this step be δ, a perturbed solution be ϑ^* and introduce

$$\vartheta = \vartheta^* + \delta$$

into (9.38); so that

$$\vartheta^* + \delta = \lambda \{R(\vartheta^*) + [(\partial/ \partial\vartheta) R(\vartheta^*)] \delta\}$$

$$+ (1-\lambda) \{R_0(\vartheta^*) + [(\partial/ \partial\vartheta) R_0(\vartheta^*)]\delta\} ,$$

or

$$C \delta = \lambda R(\vartheta^*) + (1-\lambda) R_0(\vartheta^*) - \vartheta^*$$

with

$$C = I - \lambda [(\partial/ \partial\vartheta) R(\vartheta^*)] - (1-\lambda) [(\partial/ \partial\vartheta) R_0(\vartheta^*)] . \tag{9.39}$$

The continuation of the homotopy path is easily achieved as long as the matrix C by (9.39) remains non-singular for all λ-values while ϑ^* is $\vartheta(\lambda)$.

Quite a few predictor-corrector algorithms have been proposed and most of them are based on some trial and error method; you proceed along the homotopy path as long as the iterations do converge and, otherwise, reduce your λ-step length. Suggestions for implementing a suitable approach are given by den Heijer and Rheinboldt (1980) with reference to a few numerical examples; a more accessible recent paper is due to Deuflard (1979).

When the matrix c defined by (9.39) becomes singular, then this indicates that $\vartheta(\lambda)$ is not differentiable for some $\lambda \in [0, 1]$ or, in other words, that the starting set ϑ_0 was not appropriate. What has happened?

A more involved analysis reveals that starting a continuation with some ϑ_0 may lead to a termination at some λ_1 ($\lambda_1 \le 1$), or to a discontinuity or, possibly, to a branching of the continuation path into several paths. Further, with the help of a partition of the parameter space, it is possible to count the number of out-going paths for one in-coming path entering a given part. A good account of the relevant theory can be found in Amann (1976, Chap. 3) as well as in Allgower and Georg (1980, Chap. 7); it is based on the Leray-Schauder degree for compact vector fields defined on the closure of open subsets of some Banach space.

To illustrate the type of difficulties which may occur let us consider the following very elementary situation. We are interested in the solution $\vartheta(\lambda)$ along the homotopy path defined by the two rules

$$R_0(\vartheta) = \vartheta_0 + A(\vartheta - \vartheta_0)$$

and

$$R(\vartheta) = \vartheta_1 + B(\vartheta - \vartheta_1)$$

where A and B are two arbitrary non-singular matrices. By the use of (9.38), we obtain the exact solution

$$\vartheta(\lambda) = C^{-1}[\lambda(I-B)\vartheta_1 + (1-\lambda)(I-A)\vartheta_0]$$

where the matrix

$$C = I - \lambda B - (1-\lambda)A$$

is assumed to be non-singular. When the spectra of $(I-A)$ and $(I-B)$ are

187

not appropriately similar, there exists at least one value of λ, say λ^*, which yields a singular matrix c. Then $\vartheta(\lambda^*)$ is at infinity. We see that the homotopy path can go to infinity in spite of possibly very close ϑ_0 and ϑ_1; such a behavior would lead to an algorithmic divergence. Much care is required in selecting the rules $R(\vartheta)$ and $R_0(\vartheta)$ in order to be sure that they will be in harmony.

We now restrict the above discussion to the evaluation of MM estimators. Our presentation of the Newton-Raphson solution for simultaneous equations has already provided us with a very efficient rule (9.35 - 36), namely $R(.)$ such that

$$R \begin{pmatrix} \vartheta^k \\ s^k \end{pmatrix} = \begin{pmatrix} \vartheta^{k+1} \\ s^{K+1} \end{pmatrix} = \begin{pmatrix} \vartheta^k + \delta\vartheta \\ s^k + \delta s \end{pmatrix} .$$

In what follows this rule will be denoted by its components, i. e.

$$R(.) = R(\rho, f, \vartheta, s) .$$

What initial rule $R_0(0)$ should be selected is not quite obvious. The only constraint we have met with is that the matrix c defined by (9.39) must remain non-singular along the homotopy path. However it happens that in certain situations this constraint is relatively difficult to satisfy. Therefore we are inclined to develop a more or less general method which, in essence, relies on the assumption that c has a slowly varying spectrum structure. We may also conjecture that a relatively flat ρ-function may add to the convergence difficulties because of its flatness; then substituting the square-function into the arbitrary ρ-function might be beneficial. These observations lead to the function

$$R_0(.) = \begin{pmatrix} \vartheta_2 \\ s_2 \end{pmatrix} + R_2(.)$$

with

$$R_2(.) = R(u^2, f, \vartheta, s)$$

and

$$\left(\begin{array}{c} \vartheta_2 \\ s_2 \end{array} \right) = \left(\begin{array}{c} \vartheta_0 \\ s_0 \end{array} \right) - R(u^2, f, \vartheta_0, s_0) \ .$$

On concluding this section we refer to two important points; first that there is a total freedom in the λ-parametrization and, second, that one should not trust the results blindly. The above presentation places a great deal of importance on (9.38) but this is just a possibility among several different structures. The only point of interest is the existence of a path which can be appropriately followed. Instead of (9.38) we could have written

$$\vartheta = \lambda^l R(\vartheta) + (1-\lambda^l) R_0(\vartheta)$$

for a given $l > 0$ or, a little more complex,

$$\vartheta = \lambda^l R[\lambda \vartheta + (1-\lambda) \vartheta_0] + (1-\lambda^l) R_0(\vartheta)$$

and this form might be of interest for $l \approx 0$. More generally we may consider the homotopy path relative to any parameter; for instance we can gradually vary the nominal efficiency of the ρ-function from $E = 1$ down to say $E = 0.8$. We can even imbricate into one another a scheme based on λ in (9.38) and a scheme in which the nominal efficiency is varied. We see that the flexibility is such that solutions can be obtained even in very adverse situations.

The algorithms we have described are based on the analytical properties of the equations defining the MM estimators, especially the normal equations to the minimized criteria. It should be remembered that cancellation of the derivatives does not suffice to guarantee a minimum. In non-convex situations, the validity of the solution must be thoroughly checked.

Chapter 10
Quantile estimators
and confidence intervals

We have met various techniques suitable for making point estimates in a more or less robust way; we know how to evaluate the precision of the estimators by their variance-covariance matrices and we even have the third moments. However we only have very limited information on the distributions of the estimators and can hardly draw any inference as to the sample space. Inferences on the sample space will be of first concern and then we will return to distributional aspects.

10.1. Quantile estimators

As we are (and will remain) in a difficult position when trying to assess confidence regions from our estimation procedures, it seems natural to determine the probability levels directly on the basis of the sample space. This is another view as regards the problem; instead of working into the parameter space, we always refer to the sample space.

Quite a few investigators have considered the linear regression set-up because the distribution of the residuals then depends on the parameters; with very little further complexity non-linear regressions can also be analyzed. We do not intend to review methods based on the assessment of the local density of probability in the sample space as is done through density kernels (see Devroye, 1978, or Devroye and Wagner, 1980, for instance); we will not be concerned by an overall assessment of the distribution of the residuals as made by Jaeckel (1972b) either.

The quantile estimators are multidimensional generalizations of the quantiles q_a of (7.4); the viewpoint has been developed by Hogg

(1975) for percentile regression, Yale and Forsythe (1976) for winsorized regression, completed by Griffiths and Willcox (1978) with a fully parametric model and, then, modified, extended and made attractive by Koenker and Bassett (1978). Their paper is fundamental and contains a review of all the related questions. The quantile estimators they obtain can be included in the composition of L estimators (7.5) as proposed by Ruppert and Carroll (1978, 1979 a, 1980).

Koenker and Bassett first consider the definition of the ordinary one-dimension quantile and, then, extend the definition to handle the linear regression set up. Assume we have a one-dimension sample set $\{y_1,...,y_n\}$ with empirical density of probability

$$g(y) = (\sum w_i)^{-1} \sum w_i \, \delta(y-y_i) \qquad (10.1)$$

as in (2.40); to avoid the technical difficulties presented by Dirac's function, we introduce some limiting scheme on kernels approximating the Dirac function, for instance kernel (7.3) with vanishing η. Then Koenker and Bassett observe that

the α-th sample quantile, $0 < \alpha < 1$, may be defined as any solution to the minimization problem

$$Q = \alpha \sum_k |y_k - u| + (1-\alpha) \sum_l |y_l - u| = \min \text{ for } u , \qquad (10.2)$$

$$k : y_k \geq u ,$$

$$l : y_l < u .$$

The case of the median ($\alpha = 1/2$) is, of course, well known, but the general result has languished in the status of curiosum.

As (10.2) lies at the root of the quantile estimator definition we prove that u is indeed the quantile q_α. Our presentation will be a little more general because it resolves the possible ties. We rewrite (10.2) into

$$Q = \alpha \int_u^\infty (y-u) \, g(y) dy + (1-\alpha) \int_{-\infty}^u (u-y) \, g(y) du ;$$

this, on differentiating with respect to u and transforming, becomes

$$-\alpha \int_u^\infty g(y)\, dy + (1-\alpha) \int_{-\infty}^u g(y)\, dy = 0 \,,$$

$$\alpha \left[G(u) - 1 \right] + (1-\alpha)\, G(u) = 0 \,,$$

$$G(u) = \alpha$$

and

$$u = q_\alpha = G^{-1}(\alpha) \,.$$

Substituting into u an arbitrary function of the observation leads to a solution of the general non-linear model (9.29)

$$y_i = f(z_i, \vartheta) + \varepsilon_i \,;$$

the criterion (10.2) becomes

$$Q = \alpha \sum_k |y_k - f(z_k, \vartheta)| + (1-\alpha) \sum_l |y_l - f(z_l, \vartheta)| = \min \text{ for } \vartheta \,, \tag{10.3}$$

$$k : y_k \geq f(z_k, \vartheta) \,,$$

$$l : u_l < f(z_l, \vartheta) \,.$$

Association of a probability measure to the sample space and differentiation lead to the conclusion that the solution $\hat{\vartheta}$ is such that:

a. the sample space is divided into two sets, one with negative and the other with non-negative residuals and with total weights proportional to α and $(1-\alpha)$, respectively.

b.

$$\alpha \sum_k \dot{f}(z_k, \vartheta) = (1-\alpha) \sum_l \dot{f}(z_l, \vartheta) \,. \tag{10.4}$$

A comment on (10.4) is that the solution $\hat{\vartheta}$ tends to be poorly defined for small size samples; in the case of linear models, dense sets of solutions may result for some α -values. The estimate $\hat{\vartheta}$ does not have a distribution that looks more or less multivariate normal and, thus, it does not seem very realistic to make use of its asymptotic variance-covariance matrix. The asymptotic properties have been investigated by Koenker and Bassett (1978) and the estimator efficiency is more specially studied by Ruppert and Carroll (1980). The last authors also report their results on test cases, one being the Brownlee stackloss data set; it is unfortunately not clear how to interpret their findings.

An approach to the many theoretical difficulties involved in investigating quantile estimators could be to regard them simply as M estimators based on asymmetric ρ-functions. This would imply substituting a smooth function into the absolute value function; for instance, (10.3) is equivalent to

$$\sum \rho \left[y_i - f(z_i, \vartheta) \right] = \min \text{ for } \vartheta$$

with a ρ-function such as

$$\dot{\rho}(u) = \alpha - (1 + e^{u/\eta})^{-1}$$

where η is an infinitesimal positive constant.

10.2. Confidence intervals

Relatively little information is available about the sample distributions of robust estimators. We have already given the jackknife method by which the possible bias, the variance and the symmetry (by third central moment) of their distributions can be assessed. Furthermore, we may safely conjecture that the tendency towards the normal distribution is very rapid whenever the influence function is bounded; this is studied by Boos and Serfling (1980).

This state of affairs is unsatisfactory although the question is of major practical interest. Attempts made by Huber (1968) to set confidence limits have provided some more insight into the relationships between the distribution of the robust location estimator and the distribution of the sample; but these attempts have only demonstrated the difficulty of the problem. Subsequently, he has proposed a few conjectures on studentization of robust estimates (1970); we believe that they are very reasonable. In the same line of tendency to normality, the paper on the computation of the sample distribution by Hampel

(1973b) is noteworthy, we excerpt:

> ... the third and perhaps most important point seems to be entirely new. It concerns not the question where to expand, but what to expand. Most papers consider the cumulative distribution F_n, some the density f_n, but neither approach leads to very simple expressions. It shall be argued that the most natural and simple quantity to study is the derivative of the logarithm of the density, f'_n / f_n, and this for several reasons: ...

Whether his approach is practical is not yet very clear. In certain special cases, very good power series can be obtained from the saddle-point method of Daniels. Recent progress is reported by Daniels (1976, 1980, 1983), Barndorff-Nielsen and Cox (1979), Callaert, Janssen and Veraverbeke (1980),Field (1982), Field and Hampel (1982) as well as Durbin (1980); the last paper places emphasis on the need for sufficient statistics but this should not matter too much because we assume an underlying distribution that is in agreement with our ρ-function definition.

Maybe we move in a vicious circle: we make use of robust estimators because we do not know precisely the distribution of the sample at hand, and we should know how a sample departs from a given model in order to state the distribution of the estimator. We are afraid it is fighting for a wrong cause when we try to derive theoretical distributions for robust estimators. We are afraid the only possibility is an inference from the sample itself, in spite of the evident limitations of this method. We feel that for small sample sizes it is not wise to assume any strict model whereas, for moderate or large sample sizes, models are not needed because of the tendency towards normality.

But we may also ask ourselves the question of whether we are able to estimate the mean and the scatter to be applied in studentization. The answer is positive, even for small sizes, with the help of the jackknife method. We demonstrate this (Rey, 1978, pp. 79-81) in the case of a Monte Carlo simulation where Huber estimators of location have been worked out for exponentially distributed samples of size $n=11$. This finding is supported by Maritz (1979) who claims that a sample size $n=10$ can be sufficient for the one-dimensional location-scatter problem. The very extensive simulation experiments of Shorack (1976) confirms this view while working with Hampel's estimators as long as the ρ-function has been correctly tuned; he demonstrates that a member of the Hampel's family studentizes particularly well, this member approximating a moderately trimmed mean.

We have reported rather limited results on the problem of setting confidence intervals but they have to be considered in the light of their applications. All results pertaining to the normality tendency can be immediately generalized to the multidimensional estimators; then studentization leads to Hotelling's T. A new approach to these problems is proposed by Boos (1980); he suggests to use the similarity between the s-shape one-dimension cumulative distribution and the s-shape $\dot{\rho}(u)$-function (whenever the latter is s-shaped). His preliminary results are interesting but it seems to the present author that so little is gained that is hardly justified to run the risk of using such a method.

Chapter 11
Miscellaneous

11.1. Outliers and their treatment

To some readers it may appear odd that we have not yet considered what are the outliers and how to handle them, whereas others may believe that thus far we have only been concerned with the question of how to treat them. Both attitudes are correct inasmuch as they are complementary; we have accepted any outliers and, in doing so, we have taken in account a loss of efficiency to be less dependent on them than we would be otherwise. Let us remember that our main interest is in having robust methods for comparison with non-robust standard methods. All robust methods handle outliers without having to specify whether observations are outlying or not.

A substantial body of literature has been devoted to this subject and we do not intend to review it for it is immaterial for our main topic. Let us just indicate that since Anscombe's paper (1960), various papers have tried to take into account the fact that rejecting an outlier drastically modifies the probability levels of all inferences. A survey of the most relevant problems can be found in Barnett and Lewis (1978). Specific methods have been proposed to assess the probability of an observation being an outlier; the only reliable approach seems to be based on Bonferroni's inequality (a good review is given by Doornbos, 1981, and further refinements are the concern of Walker, 1982). Identification of outliers is frequently attempted with the help of the influence function (see Draper and John for a recent account, 1981); an important drawback of the influence function is that it exhibits a sensitivity to very different features as we saw at the end of § 9.7. Let us remember that Atkinson (1980, 1981, 1982) proposes a simulatation method to assess the probability levels; his approach takes place in the broader context of the various diagnostic tools, see

Cook and Weisberg (1982).

The main feature of the robust methods is their reduced sensitivity to a departure from the assumptions. They may thus indicate unusual observations by the difference in their handlings. This has already been discussed at length and we will further illustrate the point. The analysis of two-way tables can be greatly facilitated by graphical displays and Bradu and Gabriel (1978) propose as an example of their biplot method its application to a data set of Yates (their page 54, Table 1). This data set relates to yields of cotton and consists in a seven by four table; a multiplicative model (additive with the logarithms) appears reasonable and we thus have to estimate a 10-parameters linear model ($n = 28$, $g = 10$ in (9.24)). To our surprise we found that this data set includes a very well defined outlier, the entry $y_{4,4} = 17.78$. Already a little specious while fitting a least squares linear model, the standardized residual became totally abnormal with the ρ-function Fair (6.27) at nominal efficiency $E = 0.8$. We have obtained

$$\ln(y_{4,4} / \hat{y}_{4,4}) = 4.2300\, s$$

with a scale

$$s = 0.08934;$$

for comparison let us indicate that the least squares result was

$$\ln(y_{4,4} / \hat{y}_{4,4}) = 2.7713\, s$$

with an error standard deviation

$$s = 0.09903,$$

which is hardly critical being the greatest of 28 residuals. But some doubt could remain concerning the outlying status of $y_{4,4}$. Then in order to support the finding let us consider a plot of the residuals. Fig. 9 displays all data set entries in coordinates which are the residuals of the robust fit against the residuals of the least squares ordinary method; for both sets of residuals, the multiplicative model was

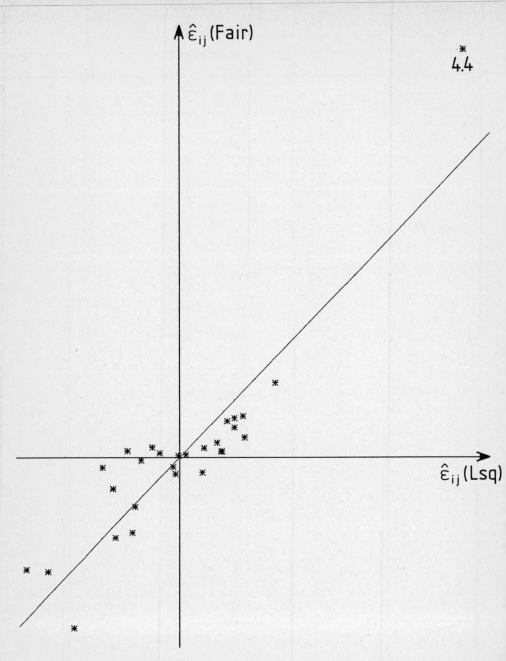

Fig.9. Yates's Analysis of Variance Data. Plot of the residuals of robust method (Fair, E = 0.8) against the least squares residuals.

assumed. It might be interesting to note that the additive model produced a similar display. Summing up, we may mark the observation $y_{4,4}$ as outlying; this may be due to a deficiency of the model or to an error in the data recording. Note that the above conclusion is original.

11.2. Analysis of variance, constraints on minimization

Considering variance analysis is equivalent to discussing the content of an empty set. There is not much to say except that this splendid tool is not trustworthy and must be redesigned from beginning to end in order to become robust. In what follows we present a few views which require a closer examination.

To set the stage, we would first like to focus our attention on an arithmetic property, namely on the decomposition of a sum of squares into two or more sums of squares. In the ordinary (least squares) set-up, the sum of squares

$$\sum t_i^2$$

can be decomposed into

$$\sum u_i^2 + \sum v_i^2$$

where u_i and v_i, the components of

$$t_i = u_i + v_i,$$

are "orthogonal". In the robust analysis of the linear model, we meet with weighted sums; the weights w_i will typically be dependent upon some residuals. We have

$$\sum w_i t_i^2 = \sum w_i (u_i + v_i)^2$$

$$= \sum w_i u_i^2 + \sum w_i v_i^2 + 2 \sum w_i u_i v_i$$

where all three terms are troublesome. The last term is a weighted cross-product with weights depending upon observations; this dependence prevents the building of "balanced" designs. It will never be possible to have strict orthogonality and therefore the computational requirements become much greater. Consider now the sums of squares. We assume that the left hand member is a variance estimate and thus we have weights which are dependent upon the t-variates, i. e.

$$\sum w_i\, t_i^2 = \sum w\,(t_i)\, t_i^2$$

We would like to find in the right hand member terms which are also variances such as

$$\sum w\,(u_i)\, u_i^2$$

but we have rather obtained ill-weighted sums, i. e.

$$\sum w\,(t_i)\, u_i^2$$

The weights differ but this could be thought of as being fairly irrelevant. Let us check the weight structure; according to (9.20) we have

$$(n-g)\, s^2 = (Int)^{-1} \sum \rho\,(t_i)$$

or

$$w\,(t_i) = (Int)^{-1} \rho\,(t_i)\, /\, t_i^2$$

where the function $\rho\,(.)$ is itself scaled with respect to s. Inasmuch as we can expect large components u_i and v_i when t_i is large, we obtain

$$w\,(t_i) \approx w\,(u_i) \approx w\,(v_i);$$

but this is not the case for outlying observations for then one of the terms, say u_i, is much too large because it relates to the residual error while the other v_i has the ordinary magnitude. We see that this weighted structure prevents the association of the sums of squares with variance effects. If nevertheless we wanted to consider these sums of squares as variances, then we should test them by the non-central F-distribution.

200

A simple (and rather reasonable) approach consists in solving the full model of linear regression, appropriately define the weights (further regarded as given a priori) and then perform the analysis of variance as it is done for any unbalanced design. There are essentially two advantages of this approach. First, since the residuals do not depend on the specific constraints introduced into parameters, the weights can be obtained in a definite way. Second, such constraints as the cancellation of the weighted sum of column effects being well written, the ordinary unbalanced analysis is feasible and we remain in the world we are familiar with. Although they are not clear on the issue, it seems that this approach does not deviate too much from the standpoint of Schrader and Hettmansperger (1980); they felt the difficulties indeed but did not know how to face them.

A rather different approach is to forget the ordinary analysis of variance and work with the full linear model (9.24) completed by (arbitrary) constraints. The numerical techniques we discussed (§§ 9.8.2, 9.8.3) admit of any number of equations and there is not much to be gained in reducing the number of parameters some what. Eventually we obtain the estimates of the effects and interactions as well as the corresponding variance-covariance matrix. This is very similar to Carroll's views (1980a), although we question the way he assesses the probability levels. With the help of the variance-covariance matrix any factor can be tested under hypothesis of, say, a student law; whether there is a row effect in a $r \times c$ table can be assessed by the Hotelling T^2 corresponding to the row factors weighted by the inverse of the appropriate submatrix of the global variance-covariance matrix and, obviously, such tests can be extended to any contrast. The interesting point in this approach is that large residuals have a tendency to inflate the scale estimate but they adversely affect neither the estimation of the parameters nor their variance-covariance estimate.

We have omitted a point of great practical importance which occurs rarely in the analysis of location-scatter problems but relatively frequently with factorial designs. In the case of factor confounding it is not feasible to define uniquely the relevant factors; this amounts to a dimensionality restriction and may cause algorithmic hazards. Program writing can be such that no trouble arises because, as can be observed, matricial inversions are never required since they are mainly convenient for an analytical description. This observation implies that generalized inverses can also be substituted into the ordinary inverses if this is preferable. Of course there is much to be said in favor of relaxing the indeterminacy, but we point out that the presented numerical procedures tolerate the indeterminacy.

11.3. Adaptive estimators

Because of the inadequacy of some estimators when applied to inappropriate distributions, many authors have tried to design sets of estimators, each member of the sets being optimal for a specific class of distributions. Then, which estimator to select for a given sample is decided either by means of tests or by setting weights. To illustrate this, we consider the location estimator

$$\vartheta = w \ (mean) + (1-w) \ (median).$$

Based on a sample $(x_1,...,x_n)$, a test may indicate the high likelihood of a heavy-tailed distribution (see Smith, 1976). The first term of the above alternative consists in setting w to zero when the test favors the heavy-tail hypothesis, and to unity if it does not; the second term of the alternative would be to define w as a monotonously increasing function of some test statistics. - A good account of the advantages and drawbacks of adaptive methods is presented by Hogg (1974), with particular emphasis on the historical background as well as the robust aspects. The paper by Takeuchi (1975) may have a broader scope although it is less informative; it is essentially a discussion of the arguments leading to robust adaptive estimators.

Frequently, fairly subjective arguments are proposed to justify the choice of the estimator structures. They range from saying that most samples are drawn from mixtures of normal distributions to declaring that man-collected data are unconsciously "corrected" to be closer to some typical value than they should. The former statement favors a location estimator intermediate between the mean and the median, whereas the latter could lead to the midrange. However, we are not very happy with the composite structure

$$\vartheta = w_1 \, (midrange) + w_2 \, (mean) + (1-w_1-w_2) \, (median).$$

Our attention has been attracted by the behavior of Tukey's outmean in Stigler's study (1977). This estimator has much in common with the midrange and it behaves reasonably well on the (man-collected) data sets; this observation supports the feeling that Stigler's data are abnormally distributed, this author suspects that samples are drawn from trimmed long-tail distributions, (the trimming being done to preserve "good" quality). Whatever the reason, Hogg justifies the poor behavior of his proposal "It seems that each of us had on his "Cauchy-colored" glasses and tried too hard to protect against outliers" (Discussion of Stigler's paper, p. 1089).

Most of the work has been oriented towards the location-scatter problem and the involved methods try to ascertain which distribution family among several very different has produced the sample. The methods are often highly heuristic (see Moberg, Ramberg and Randles, 1980); they may be based on very impressive simulation experiments and yield unexpected lessons. We find in Harter, Moore and Curry (1978, p. 25)

> The Monte Carlo results show one rather surprising phenomenon. Some of the relative efficiencies of the adaptive robust estimators are larger than 1 (100 %)....
> Another possible explanation is that if a sample from population A behaves more like a sample from population B, it may actually be more efficient to use the estimator appropriate for population B, which would be the adaptive robust estimator.

The situation is not much different when the various estimator components are drawn from a common class. Yohai (1974) suggests to select the "best" estimator among a family of Huber's estimators corresponding to various parameters k and proves the asymptotic optimality of the method.

In this context, we would rather prefer the bayesian viewpoint because this makes it clear what is the origin of the arbitrariness. An involved treatment is proposed by Miké (1973) who introduces a family of prior distributions; but possibly the soundest treatment would be to use prior distributions only defined by upper and lower bounds. This is in line with Dempster (1968) as well as Walley and Fine (1982), but this author has not heard of any significant application in the robustness field. It appears important to refrain from inferring too much when the sample provides only little evidence.

To conclude these mixed feelings, there is the comforting remark made by Relles and Rogers (1977) : statisticians are fairly robust estimators of location.

11.4. Recursive estimators

Particularly in time-series analysis, it is peremptory to have expressions which permit an estimator to be evaluated from a sample of size $(n + 1)$, when the estimator based on size n as well as some additional information are given. Makhoul (1975) surveys the various methods relevant to forecasting with linear models and they appear to be non-robust.

The design of robust methods has, so far, been unsatisfactory. Some theory for recursive estimators of type M has been developed by Nevel'son (1975); it presents similarities with the Robbins-Monro algorithm and relates to one-step M estimators. Whether the finite sample properties are satisfactory or not remains an open question.

With regard to filtering, Tollet (1976) proposes a Kalman-Bucy approach with a gain factor selected in order to keep the sensitivity to gross errors or outliers within bounds. Occasional outliers have a weak influence, but this influence is not strictly bounded in order to permit tracking in the presence of a large offset. A more accessible paper by Masreliez and Martin (1977) is partially similar. Approximately the same strategy had been proposed by Rey (1974) with a recursive estimation of the "median" of serial data. His work also deals with a robust estimation of the scale analogous to Kersten and Kurz's proposal (1976), based on the paper by Evans et al. (1976). On similar lines we should mention the work of Safanov and Athans (1978) as well as that of Price and Vandelinde (1979) who demonstrate that the research activity is wealthy. A different approach to these problems consists in the attempts to smooth rather than filter the data flow; a "local" treatment is preferred over a processing that depends upon all the past information. Many different techniques are presented by Gasser and Rosemblatt (1979) in a collection of papers prepared for a workshop; most contributions deserve the greatest attention and we recommend the first two pages of Martin (1979) for a well arranged survey of the recent references.

It does not seem that robust estimation is considered in the investigation of point processes, except by few people. Gaver and Hoel (1970) have compared several estimates of a reliability factor in a Poisson process, placing emphasis on bias, variance and sensitivity to Poisson hypothesis. They conclude that an estimate derived from the jackknife theory is optimal in most respects. This is supported by a fundamental study due to Marcus and Woo (1979).

In time-series as well as in point process analyses, the limited results can be attributed to the fact that it is difficult to assess what are the key assumptions of the models and to evaluate the sensitivity to these assumptions. Among them, the independency assumption is almost unmanageable: it is frequently opposed to linear correlation, but this is only one of the possible alternate hypotheses. This question is developed with the help of an influence function by Devlin et al. (1975) and by Mallows (1975); the results are too complicate to be very promising. It is not even very clear sometimes what type of robustness is required, which induced Papantoni-Kazakos (1980) to suggest the use of the Vasershtein distance rather than the Prokhorov distance. Papantoni-Kazakos and Kazakos (1979) feel that the distance sensitivity to the extreme offsets (outliers) is too great and propose a measure that is directly related to the average cost of an estimation error. At first sight their approach seems sensitive to independency conditions.

11.5. Concluding remark

Let us recall that in data analysis robust estimation is, to us, an essential complement to the more classical methods. Robust methods help us to validate computation results by providing a reliable basis of comparison. The need for these methods is particularly evident when multivariate data are to be processed. In routine use of robust, quasi-robust and more or less informal methods, we have uncovered many features in data sets which were concealed in the analysis results; this was found from a comparison between similar treatments that only differed in robustness.

Chapter 12
References

Abramov V. A., Estimates for the Levy-Prokhorov distance, Theory Probab. Appl. **21** (1976) 396-400.

Aitkin M. and Wilson G. T., Mixture models, outliers, and the EM algorithm, Technometrics, **21** (1980) 325-331.

Allgower E. and Georg K., Simplicial and continuation methods for approximating fixed points and solutions to system of equations, SIAM Rev. **22** (1980) 28-85.

Amann H., Fixed point equations and non-linear eigenvalue problems in ordered Banach spaces, SIAM Rev. **18** (1976) 620-709.

Andrews D. F., A general method for the approximation of tail areas, Ann. Statist. **1** (1973) 367-372.

Andrews D. F., A robust method for multiple linear regression, Technometrics, **16** (1974) 523-531.

Andrews D. F., Bickel P. J., Hampel F. R., Huber P. J., Rogers W. H. and Tukey J. W., Robust Estimates of Location: Survey and Advances, Princeton Univ. Press, Princeton, (1972).

Andrews D. F. and Pregibon D., Finding outliers that matter, J. Roy. Statist. Soc. **B-40** (1978) 85-93.

Anscombe F. J., Rejection of outliers, Technometrics, **2** (1960) 123-147.

Atkinson A. C., Examples showing the use of two graphical displays for the detection of influential and outlying observation in regression, COMPSTAT 1980, (1980) 276-282.

Atkinson A. C., Two graphical displays for outlying and influential observations in regression, Biometrika, **68** (1981) 13-20.

Atkinson A. C., Regression diagnostics, transformations and constructed variables, J. Roy. Statist. Soc. **B-44** (1982) 1-36.

Balinski M. L. and Cottle R. W. (Eds.), Complementarity and Fixed Point Problems, North-Holland, (1978).

Bard Y., Nonlinear Parameter Estimation, Academic Press, (1974).

Barndorff-Nielsen O. and Cox D. R., Edgeworth and saddle-point approximations with statistical applications, J. Roy. Statist. Soc. **B-41** (1979) 279-312.

Barnett V. and Lewis T., Outliers in Statistical Data, Wiley, (1978).

Barra J. R., Brodeau F., Romier G. and Van Cutsem B. (Eds.), Recent Developments in Statistics, North-Holland, (1977).

Bates D. M. and Watts D. G., Relative curvature measures of nonlinearity, J. Roy. Statist. Soc. **B-42** (1980) 1-25.

Beale E. M. L., Confidence regions in non-linear estimation, J. Roy. Statist. Soc. **B-22** (1960) 41-88.

Beaton A. E. and Tukey J. W., The fitting of power series, meaning polynomials, illustrated on band spectroscopic data, Technometrics, **16** (1974) 147-185.

Beran R., Robust location estimates. Ann. Statist. **5** (1977a) 431-444.

Beran R., Minimum Hellinger distance estimates for parametric models. Ann. Statist. **5** (1977b) 445-463.

Beran R., Robust estimations in models for independent non-identically distributed data, Ann. Statist. **10** (1982) 415-428.

Berger J. O., Inadmissibility results for generalized bayes estimators of coordinates of a location vector, Ann. Statist. **4** (1976a) 302-333.

Berger J. O., Admissibility results for generalized bayes estimators of coordinates of a location vector, Ann. Statist. **4** (1976b) 334-356.

Berkson J., Minimum X-square, not maximum likelihood, Ann. Statist. **8** (1980) 457-487.

Bickel P. J., One-step Huber estimates in the linear model, J. Amer. Statist. Ass. **70** (1975) 428-434.

Bickel P. J., Using residuals robustly; I: Tests for heteroscedasticity, nonlinearity, Ann. Statist. **6** (1978) 266-291.

Bickel P. J. and Freedman D. A., Some asymptotic theory for the bootstrap, Ann. Statist. **9** (1981) 1196-1217.

Bierens H. J., Robust methods and asymptotic theory in nonlinear econometrics, Ph. D. Dissertation, Amsterdam Univ., Amsterdam, (1980).

Bierens H. J., Robust methods and asymptotic theory in nonlinear econometrics, Lecture Notes Econ. Math. Syst., Springer Verlag, **192** (1981).

Birch J. B., Effects of the starting value and stopping rule on robust estimates obtained by iterated least squares, Commun. Statist. - Simula. Computa. **9** (1980) 141-154.

Bissel A. F. and Ferguson R. A., The jackknife: Toy, tool or two edged weapon? Statistician, **24** (1975) 79-100.

Boos D. D., A differential for L statistics, Ann. Statist. **7** (1979) 955-959.

Boos D. D., A new method for constructing approximate confidence intervals from M estimates, J. Amer. Statist. Ass. **75** (1980) 142-145.

Boos D. D. and Serfling R. J., A note on differentials and the CLT and LIL for statistical functions with application to M estimates, Ann. Statist. **8** (1980) 618-624.

Box G. E. P., Non-normality and test on variances, Biometrika, **40** (1953) 318-335.

Boyd D. W., The power method for Lp norms, Lin. Alg. and its Appl., **9** (1974) 95-101.

√ Boyer J. E. and Kolson J. O., Variances for adaptive trimmed means, Biometrika, **70** (1983) 97-102.

Bradu D. and Gabriel K. R., The biplot as a diagnostic tool for model of twoway tables, Technometrics, **20** (1978) 47-68.

Brownlee K. A., Statistical Theory and Methodology in Science and Engineering, Wiley, 2nd ed. (1965).

Callaert H., Janssen P. and Veraverbeke N., An Edgeworth expansion for U-statistics, Ann. Statist. **8** (1980) 299-312.

Campbell N. A., Robust procedures in multivariate analysis, I. Robust covariance estimation, Appl. Statist. **29** (1980) 231-237.

Campbell N. A., Robust procedures in multivariate analysis, II. Robust canonical variate analysis, Appl. Statist. **31** (1982) 1-8.

Cargo G. T. and Shisha O., Least pth powers of deviations. J. Approxim. Theory, **15** (1975) 335-355.

Carroll R. J., On the asymptotic distribution of multivariate M estimates, J. Multiv. Anal. **8** (1978) 361-371.

Carroll R. J., Robust methods for factorial experiments with outliers, Appl. Statist. **29** (1980a) 246-251.

Carroll R. J., Robust estimation in the heteroscedastic model when there are many parameters, Univ. North Carolina, Chapel Hill, **1321** (1980b).

Carroll R. J. and Ruppert D., On robust tests for heteroscedasticity, Univ. North Carolina, Chapel Hill, **1241** (1979).

√ Chambers R. L. and Heathcote C. R., On the estimation of slope and the identification of outliers in linear regression, Biometrika, **68** (1981) 21-

33.

Chan L. K. and Rhodin L. S., Robust estimation of location using optimally chosen sample quantiles, Technometrics, **22** (1980) 225-237.

Chen C. H., On information and distance measures, error bounds and feature selection, Information Sc. **10** (1976) 159-174.

Clarke G. P. Y., Moments of the least squares estimators in non-linear regression model, J. Roy. Statist. Soc. **B-42** (1980) 227-237.

Coleman D., Holland P., Kaden N., Klema V. and Peters S. C., A system of subroutines for iteratively reweighted least squares computations, ACM Trans. Math. Softw. **6** (1980) 327-336.

Collins J. R., Robust estimation of a location parameter in the presence of asymmetry, Ann. Statist. **4** (1976) 68-85.

Collins J. R., Robust M estimators of location vectors, J. Multiv. Anal. **12** (1982) 480-492.

Conover W. J., Rank tests for one sample, two samples and K samples without the assumption of a continuous distribution function, Ann. Statist. **1** (1973) 1105-1125.

Cook R. D. and Weisberg S., Residuals and Influence in Regression, Chapman and Hall, London (1982).

Cressie N., Transformations and the jackknife, J. Roy. Statist. Soc. **B-43** (1981) 177-182.

Cryer J. D., Robertson T. and Wright F. T., Monotone median regression, Ann. Math. Statist. **43** (1972) 1459-1469.

Daniel C. and Wood F. S., Fitting Equations to Data, Wiley, (1971).

Daniels H. E., Approximation to the distribution of Huber's robust M estimators (Abstract), European Meeting of Statisticians, Université de Grenoble (1976) 180.

Daniels H. E., Saddlepoint approximations for estimating equations, Biometrika, **70** (1983) 89-96.

Das Gupta S., Brunn-Minkowski inequality and its aftermath, J. Multiv. Anal. **10** (1980) 296-318.

Daniels H. E., Exact and saddlepoint approximations, Biometrika, **67** (1980) 59-63.

David H. A. (Ed.), Contributions to survey Sampling and Applied Statistics, Academic Press (1978).

David H. A. and Shu V. S., Robustness of location estimators in the presence of an outlier, in David (1978) 235-250.

Dempster A. P., Estimation in multivariate analysis, in Krishnaiah (1966) 315-334.

Dempster A. P., A generalization of bayesian inference, J. Roy. Statist. Soc. **B-30** (1968) 205-247.

Dempster A. P., A subjectivist look at robustness, in ISI, **1** (1975) 349-374.

Dempster A. P., Examples relevant to the robustness of applied inferences, in Gupta and Moore (1977) 121-138.

Denby L. and Mallows C. L., Two diagnostic displays for robust regression analysis, Technometrics, **19** (1977) 1-13.

den Heijer C. and Rheinboldt W. C., On steplength algorithms for a class of continuation methods, Univ. Pittsburgh, Pittsburgh, Techn. Rept. **ICMA-80-16** (1980).

Dennis J. E. and Welsch R. E., Techniques for nonlinear least squares and robust regression, Proc. Amer. Statist. Ass., (1976) 83-87.

Deuflhard P., A stepsize control for continuation methods and its special application to multiple shooting techniques, Numer. Math. **33**

(1979) 115-146.

Devlin S. J., Gnanadesikan R. and Kettenring J. R., Robust estimation of dispersion matrices and principal components, J. Amer. Statist. Ass. **76** (1981) 354-362.

Devroye L. P., The uniform convergence of the Nadaraya-Watson regression function estimate, Canad. J. Statist. **6** (1978) 179-191.

Devroye L. P. and Wagner T. J., Distribution-free consistency results in nonparametric discrimination and regression function estimation, Ann. Statist. **8** (1980) 231-239.

Diaconis P. and Efron B., Computer-intensive methods in statistics, Scientific Amer. (May 1983) 96-108.

Doornbos R., Testing for an outlier in a linear model, Biometrics, **37** (1981) 705-711.

Dowson D. C. and Landau B. V., The Frechet distance between multivariate normal distributions, J. Multiv. Anal. **12** (1982) 450-455.

Draper N. R. and John J. A., Influential observations and outliers in regression, Technometrics, **23** (1981) 21-26.

Draper N. R. and Smith H., Applied Regression Analysis, Wiley, (1966).

Durbin J., Approximations for densities of sufficient estimators, Biometrika, **67** (1980) 311-333.

Dutter R., Algorithms for the Huber estimator of multiple regression, Computing, **18** (1977) 167-176.

Dutter R. and Huber P. J., On the methods for the numerical solution of robust regression problems, Manuscript, Technical Univ. Graz, (1978).

Dutter R. and Huber P. J., Numerical methods for the nonlinear robust regression problem, J. Statist. Comput. Simul. **13** (1981) 79-113.

Eaton M. L., A review of selected topics in multivariate probability inequalities, Ann. Statist. **10** (1982) 11-43.

Edgeworth F. Y., Exercises in the calculation of errors, Philos. Magaz. **36** (1893) 98-111.

Efron B., Bootstrap methods: Another look at the jackknife, Ann. Statist. **7** (1979a) 1-26.

Efron B., Computers and the theory of statistics: Thinking the unthinkable, SIAM Rev. **21** (1979b) 460-480.

Efron B., The jackknife, the bootstrap and other resampling plans, Stanford Univ., Stanford, Techn. Rept. **163** (1980b).

Efron B., Nonparametric estimates of standard errors: The jackknife, the bootstrap and other methods, Biometrika, **68** (1981a) 589-599.

Efron B., Censored data and the bootstrap, J. Amer. Statist. Ass. **76** (1981b) 312-319.

Efron B. and Hinkley D. V., Assessing the accuracy of the maximum likelihood estimator: observed versus expected Fisher information, Biometrika, **65** (1978) 457-487.

Eisenhart C., The development of the concept of the best mean of a set of measurements from antiquity to the present day, Amer. Statist. Ass. Presidential Address, Audience notes (1971).

Ekblom H., Calculation of linear best Lp-approximations, BIT, **13** (1973) 292-300.

Ekblom H., Lp-methods for robust regression, BIT, **14** (1974) 22-32.

Evans J. G., Kersten P. and Kurz L., Robust recursive estimation with applications, Inform. Sc. **11** (1976) 69-92.

Feller W., An Introduction to Probability Theory and its Applications, Wiley, Vol.2 (1966).

Fenstad G. U., Kjaernes M. and Walloe L., Robust estimation of standard deviation, J. Comput. Simul. **10** (1980) 113-132.

Ferguson T. S., Prior distributions on spaces of probability measures, Ann. Statist. **2** (1974) 615-629.

Ferguson R. A., Fryer J. G. and Mc Whinney I. A., On the estimation of a truncated normal distribution, in I. S. I. **3** (1975) 259-363.

Field C., Small sample asymptotic expansions for multivariate M estimates, Ann. Statist. **10** (1982) 672-689.

Field C. A. and Hampel F. R., Small asymptotic distributions of M estimators of location, Biometrika, **68** (1982) 29-46.

Fine T. L., Theories of Probability: An Examination of Foundations, Academic Press, (1973).

Fischler M. A. and Bolles R. C., Random sample consensus: A paradigm for model fitting with applications to image analysis and automated cartography, Comm. ACM, **24** (1981) 381-395.

Fisher R. A., On the mathematical foundations of theoretical statistics, Philos. Trans. Roy. Soc. London, **A-222** (1922) 309-368.

Fletcher R., Grant J. A. and Hebden M. D., The continuity and differentiability of the parameters of the best linear Lp-approximation, J. Approx. Theory, **10** (1974) 69-73.

Fletcher R. and Powell M. J. D., A rapidly convergent descent method for minimization, Computer J. **6** (1963) 163-168.

Forster W. (Ed.), Numerical Solution of Highly Nonlinear Problems, North-Holland, (1980).

Forsythe A. B., Robust estimation of straight line regression coefficients by minimizing pth power deviations, Technometrics, **14** (1972) 159-166.

Freedman D. A., Bootstrapping regression models, Ann. Statist. **9** (1981) 1218-1228.

Fuller W. A., Properties of some estimators for the errors-in variables model, Ann. Statist. **8** (1980) 407-422.

Garber S. and Klepper S., Extending the classical errors-in-variables model, Econometrica, **48** (1980) 1541-1546.

Garcia C. B. and Gould F. J., Relations between several path following algorithms and local and global Newton methods, SIAM Rev. **22** (1980) 263-274.

Garcia-Polamores U. and Giné E. M., On the linear programming approach to the optimality property of Prokhorov's distance, J. Math. Anal. and Appl. **60** (1977) 596-600.

Gasser T. and Rosenblatt M. (Eds.), Smoothing Techniques for Curve Estimation, Lecture Notes in Mathematics, Springer-Verlag, **757** (1979).

Gaver D. P. and Hoel D. G., Comparison of certain small sample Poisson probability estimates, Technometrics, **12** (1970) 835-850.

Gentleman W. M., Robust estimation of multivariate location by minimizing pth power deviations, Ph. D. Dissertation, Princeton University, and Bell Tel. Labs, Memorandum **MM65-1215-16** (1965).

Ghosh M. and Sinha B. K., On the robustness of least squares procedures in the regression models, J. Multiv. Anal. **10** (1980) 332-342.

Gill P. E. and Murray W., Algorithms for the solution of the non-linear least-squares problem, SIAM J. Numer. Anal. **15** (1978) 977-992.

Gleser L. J., Estimation in a "errors in variables" regression model: Large sample results, Ann. Statist. **9** (1981) 24-44.

Golub G. E. and van Loan C. F., An analysis of the total least squares problem, SIAM J. Numer. Anal. **17** (1980) 883-893.

Gray H. L. and Schucany W. R., The Generalized Jackknife Statistic, Marcel Dekker, (1972).

Gray H. L., Schucany W. R. and Watkins T. A., On the generalized jackknife and its relation to statistical differentials, Biometrika, **62** (1975) 637-642.

Gray H. L., Schucany W. R. and Woodward W. A., Best estimates of functions of the parameters of the gaussian and gamma distributions. IEEE Trans. Reliab. **R-25** (1976) 95-99.

Griffiths D. and Willcox M., Percentile regression: A parametric approach, J. Amer. Statist. Ass. **73** (1978) 496-498.

Gross A. J. and Hosmer D. W., Approximating tail areas of probability distributions, Ann. Statist. **6** (1978) 1352-1359.

Gross A. M., Confidence intervals for bisquare regression estimates, J. Amer. Statist. Ass. **72** (1977) 341-354.

Gross A. M. and Tukey J. W., The estimators of the Princeton Robustness Study, Princeton Univ. Techn. Rept. **38-2** (1973).

Gupta S. S. and Moore D. S. (Eds.), Statistical Decision Theory and Related Topics, Academic Press, (1977).

Hall P., On the limiting behavior of the mode and median of the sum of independent random variables, Ann. Prob. **8** (1980) 419-430.

Hampel F. R., Contributions to the theory of robust estimation, Ph. D. Dissertation, Univ. California, Berkeley, (1968).

Hampel F. R., A general qualitative definition of robustness, Ann. Math. Statist. **42** (1971) 1887-1896.

Hampel F. R., Robust estimation: a condensed partial survey, Z. Wahrscheinlichkeitstheorie verw. Geb. **27** (1973a) 87-104.

Hampel F. R., Some small sample asymptotics, Proc. Prague Symposium on asymptotic statistics (1973b) 109-126.

Hampel F. R., The influence curve and its role in robust estimation, J. Amer. Statist. Ass. **69** (1974) 383-393.

Hampel F. R., Beyond location parameters: Robust concepts and methods, In I. S. I. **1** (1975) 375-382.

Hampel F. R., On the breakdown points of some rejection rules with mean, E. T. H. Zurich, Res. Rept. **11** (1976).

Hampel F. R., Modern trends in the theory of robustness, Math. Operationsforsch. Statist. - Ser. Statist. **9** (1978a) 425-442.

Hampel F. R., Optimally bounding the gross-error sensitivity and the influence of position in factor space, ASA/IMS Meeting, San Diego (1978b).

Hampel F. R., Marazzi A., Ronchetti E., Rousseeuw P., Stahel W. and Welsch R. E., Bibliography, Handouts for the instructional meeting on Robust Statistical Methods, 15th European Meeting of Statisticians, Palermo, Sept. 13-17, **5** (1982).

Hampel F. R., Rousseeuw P. and Ronchetti E., The change-of-variance curve and optimal redescending M estimators, J. Amer. Statist. Ass. **76** (1981) 643-648.

Harter H. L., The method of least squares and some alternatives, Int. Statist. Rev. **42** (1974) 147-174, 235-264, 282, **43** (1975) 1-44, 125-190, 273-278, 269-272, **44** (1976) 113-159.

Harter H. L., Moore A. H. and Curry T. F., Adaptive robust estimation of location and scale parameters of symmetric populations, Air Force Flight Dynamic Laboratory, Wright-Patterson Air Force Base, AFFDL-TR-78-128 (1978).

Hartigan J. A., Necessary and sufficient conditions for asymptotic joint normality of a statistic and its subsample values, Ann. Statist. **3** (1975) 573-580.

Harvey A. C., A comparison of preliminary estimators for robust regression, J. Amer. Statist. Ass. **72** (1977) 910-913.

Henrici P., Applied and Computational Complex Analysis, Wiley, Vol. 1 (1974).

Hettmansperger T. P. and Mc Kean J. W., A robust alternative based on ranks to least squares in analyzing linear models, Technometrics, **19** (1977) 275-284.

Hiebert K. L., An evaluation of the mathematical software that solves non-linear least squares problems, ACM Trans. Math. Softw. **7** (1981) 1-16.

Hill R. W., Robust regression when there are outliers in the carriers, Ph. D. Dissertation, Harvard Univ., Boston (1977).

Hill R. W., On estimating the covariance matrix of robust regression M estimates, Commun. Statist. **A8** (1979) 1183-1196.

Hill R. W. and Holland P. W., Two robust alternatives to least-squares regression, J. Amer. Statist. Ass. **72** (1977) 828-833.

Hinkley D. and Wang H. L., A trimmed jackknife, J. Roy. Statist. Soc. **B-42** (1980) 347-356.

Hogg R. V., Adaptive robust procedures: A partial review and some suggestions for future applications and theory, J. Amer. Statist. Ass. **69** (1974) 909-927.

Hogg R. V., Estimates of percentile regression lines using salary data, J. Amer. Statist. Ass. **70** (1975) 56-59.

Hogg R. V., Statistical robustness: One view of its use in applications today, Amer. Statistician, **33** (1979a) 108-115.

Hogg R. V., An introduction to robust estimation, in Launer and Wilkinson (1979b) 1-17.

Huber P. J., Robust estimation of a location parameter, Ann. Math. Statist. **35** (1964) 73-101.

Huber P. J., Robust confidence limits, Z. Wahrscheinlichkeitstheorie verw. Geb. **10** (1968) 269-278.

Huber P. J., Théorie de l'inférence statistique robuste, Presses Univ., Montréal, (1969).

Huber P. J., Studentizing robust estimates, in Puri (1970) 453-463.

Huber P. J., Robust statistics: A review, Ann. Math. Statist. **43** (1972) 1041-1067.

Huber P. J., Robust regression: asymptotics, conjectures and Monte Carlo, Ann. Statist. **1** (1973) 799-821.

Huber P. J., Robust covariances, in Gupta and Moore (1977a) 165-191.

Huber P. J., Robust methods of estimation of regression coefficients, Math. Operationsforsch. Statist. - Ser. Statist. **8** (1977b) 41-53.

Huber P. J., Robust Statistical Procedures, SIAM monograph, Philadelphia, **27** (1977c).

Huber P. J., Robust Statistics, Wiley (1981).

Huber P. J., Minimax aspects of bounded-influence regression, J. Amer. Statist. Ass. **78** (1983) 66-80.

Huber P. J. and Dutter R., Numerical solution of robust regression problems, COMPSTAT 1974, (1974) 165-172.

Hwang S. Y., On monotonicity of Lp and lp norms, IEEE Trans. Electr. Acoust. Speech and Sign. Proc. **ASSP-23** (1975) 593-594.

I. S. I., Proc. 40th Session Int. Statist. Inst., Warsaw-1975, 4 books, Bull. Int. Statist. Inst., **46** (1975).

I. S. I., Proc. 42th Session Int. Statist. Inst., Philippines-1979, 4 books, Bull. Int. Statist. Inst., **48** (1979).

Jackson D., Note on the median of a set of numbers, Bull. Amer. Math. Soc. **27** (1921) 160-164.

Jaeckel L. A., Robust estimates of location: Symmetry and asymmetric contamination, Ann. Math. Statist. **42** (1971) 1020-1034.

Jaeckel L. A., The infinitesimal jackknife, Bell Tel. Labs, Memorandum **MH-1215** (1972a).

Jaeckel L. A., Estimating regression coefficients by minimizing the dispersion of residuals. Ann. Math. Statist. **43** (1972b) 1449-1458.

James B. R. and James K. L., Analogues of R estimators for quantal bioassay, Stanford Univ., Stanford, Techn. Rept. **49** (1979).

Jewett R. I. and Ronner A. E., Trimmed means and M estimates, Statistica Neerl. **35** (1981) 221-227.

Johnson D. E., Mc Guire S. A. and Milliken G. A., Estimating σ^2 in the presence of outliers, Technometrics, **20** (1978) 441-445.

Johnson R. W., Axiomatic characterization of the directed divergences and their linear combinations, IEEE Trans. Inform. Theory, **IT-25** (1979) 709-716.

Kanal L., Patterns in pattern recognition, IEEE Trans. Information Theory, **IT-20** (1974) 697-722.

Karamardian S. (Ed.), Fixed Points Algorithms and Applications, Academic Press (1977).

Kellog R. B., Li T. Y. and Yorke J., A constructive proof of the Brouwer fixed point theorem and computational results, SIAM J. Numer. Anal.

13 (1976) 473-483.

Kendall M. G. and Buckland W. R., A Dictionary of Statistical Terms (prepared for Int. Statist. Inst.), Oliver and Boyd, Edinburgh, 4th ed. (1981).

Kersten P. and Kurz L., Robustized vector Robbins-Monro algorithm with applications to M interval detection, Information Sc. **11** (1976) 121-140.

Khatri C. G., Unified treatment of Cramer-Rao bound for the nonregular density functions, J. Statist. Plann. Inform. **4** (1980) 75-79.

Klema V., Rosepack - Robust estimation package, ACM Signum letter, **13** (1978) 18-19.

Koenker R. And Bassett G., Regression quantiles, Econometrica, **46** (1978) 33-50.

Komlos J., Major P. and Tusnady G., Weak convergence and embedding, in Revesz (1975) 149-165.

Kozek A., Efficiency and Cramer-Rao type inequalities for convex loss functions, J. Multiv. Anal. **7** (1977) 89-106.

Krasker W. S., Estimation in linear regression models with disparate data points, Econometrica, **48** (1980) 1333-1346.

Krasker W. S. and Welsch R. E., Efficient bounded-influence regression estimation, J. Amer. Statist. Ass. **77** (1982) 595-604.

Krishnaiah P. R. (Ed.), Multivariate Analysis, Academic Press, Vol. 1 (1966).

Kubat P., Mean or median? (a note on an old problem), Statistica Neerlandica, **33** (1979) 191-196.

Lachenbruck P. A., On expected probabilities of misclassification in discriminant analysis, necessary sample size, and a relation with the multiple correlation coefficient, Biometrics, **24** (1968) 823-834.

Lambert D., Influence functions for testing, J. Amer. Statist. Ass. **76** (1981) 649-657.

Landers D. and Rogge L., On nonuniform gaussian approximation for random simulation, Preprints in Statistics, Univ. Cologne, **22** (1976).

Launer R. L. and Wilkinson G. N. (Eds.), Robustness in Statistics, Academic Press (1979).

Lenth R. V., A computational procedure for robust multiple regression, Univ. Iowa, Iowa City, Techn. Rept. **53** (1976).

Leprêtre L., Lois limites du jackknife de statistiques associées à des fonctions dérivables au sens de von Mises, Rev. Statist. Appliquées, **27** (1979) 55-79.

Lewis J. T. and Shisha O., Lp-convergence of monotone functions and their uniform convergence, J. Approxim. Theory, **14** (1975) 281-284.

Major P., On the invariance principle for sums of independent identically distributed random variables, J. Multiv. Anal. **8** (1978) 487-517.

Makhoul J., Linear prediction: A tutorial review, Proc. IEEE, **63** (1975) 693-708.

Malinvaud E., Statistical Methods of Econometrics, North_Holland, 2nd ed. (1970).

Mallows C. L., On some topics in robustness, I. M. S. meeting, Rochester (1975).

Marcus A. H. and Woo J., Robust estimates of reliability in small samples: Part 1, theory, Washington State University, Pullman (1979).

Maritz J. S., A note on exact robust confidence intervals for location, Biometrika, **66** (1979) 163-166.

Maritz J. S. and Jarrett R. G., A note on estimating the variance of the sample median, J. Amer. Statist. Ass. **73** (1978) 194-196.

Maronna R. A., Robust M estimators of multivariate location and scatter, Ann. Statist. **4** (1976) 51-67.

Maronna R., Bustos O. and Yohai V., Bias- and efficiency-robustness of general M estimator for regression with random carriers, in Gasser and Rosenblatt (1979) 91-116.

Martin R. D., Approximate conditional-mean type smoothers and interpolators, in Gasser and Rosenblatt (1979) 117-143.

Mason D. M., Asymptotic normality of linear combinations of order statistics with a smooth score function, Ann. Statist. **9** (1981) 899-908.

Masreliez C. J. and Martin R. D., Robust bayesian estimation for the linear model and robustifying the Kalman filter, IEEE Trans. Autom. Control, **AC-22** (1977) 361-371.

Mc Kean J. W. and Schrader R. M., The geometry of robust procedures in linear models, J. Roy. Statist. Soc. **B-42** (1980) 366-371.

Mead R. and Pike D. J., A review of response surface methodology from a biometric viewpoint, Biometrics, **31** (1975) 803-851.

Merle G. and Spath H., Computational experiences with discrete Lp approximation, Computing, **12** (1974) 315-321.

Merriam-Webster A., Webster's New Ideal Dictionary, G. and C. Merriam Cy., Publishers, Springfield, Mass.

Miké V., Robust Pitman-type estimators of location, Ann. Inst. Statist. Math. **25** (1973) 65-86.

Miller R. G., The Jackknife: A review, Biometrika, **61** (1974) 1-15.

Miller R. G. and Halpern J. W., Robust estimators for quantal bioassay, Biometrika, **67** (1980) 103-110.

Moberg T. F., Ramberg J. S. and Randles R. H., An adaptive multiple regression procedure based on M estimators, Technometrics, **22** (1980) 213-224.

Morineau A., Régressions robustes, méthodes d'ajustement et de validation, Rev. de Statist. Appl. **26** (1978) 5-28.

Mosteller F., The jackknife, Rev. Int. Statist. Inst. **39** (1971) 363-368.

Mosteller F., Rourke R. E. K. and Thomas G. B., Probability with Statistical Applications, Addison-Wesley, 2nd ed. (1970).

Munster M., Théorie générale de la mesure et de l'intégration, Bull. Soc. Roy. Sc. Liège, **43** (1974) 526-567.

Nakajima F. and Kozin F., A characterization of consistent estimators, IEEE Trans. Autom. Contr. **AC-24** (1979) 758-765.

N. C. H. S., Annotated bibliography on robustness studies of statistical procedures, U. S. Dept. Health Educ. Welf., Publication (HSM) **72-1051** (1972).

Nelson W., Theory and applications of hazard plotting for censored failure data, Technometrics, **14** (1972) 945-966.

Nevel'son M. B., On the properties of the recursive estimates for a functional of an unknown distribution function, in Revesz (1975) 227-251.

Nirenberg L., Variational and topological methods in nonlinear problems, Bull. Amer. Math. Soc. **4** (1981) 267-302.

Olkin I. (Ed.), Contributions to Probability and Statistics, Stanford Univ. Press, (1960).

Ortega J. M. and Rheinboldt W. C., Iterative Solution of Nonlinear Equations in Several Variables, Academic Press, (1970).

Papantoni-Kazakos P., Robustness in parameter estimation, IEEE Trans. Information Theory, IT-23 (1977) 223-231.

Papantoni-Kazakos P., The Vasershtein distance as the stability criterion in robust estimation, IEEE Trans. Inform. Theory, IT-26 (1980) 620-625.

Papantoni-Kazakos P. and Kazakos D., Nonparametric Methods in Communication, Electrical Engineering and Electronics, Marcel Dekker, (1977).

Parr W. C. and Schucany W. R., The jackknife bibliography, Int. Statist. Rev. 48 (1980) 73-78.

Parr W. C. and Schucany W. R., Jackknifing L statistics with smooth weight functions, J. Amer. Statist. Ass. 77 (1982) 629-638.

Pitman E. J. G., The estimation of location and scale parameters of a continuous population of any given form. Biometrika, 30 (1939) 391-421.

Poincare H., Calcul des Probabilités, Gauthier-Villars, (1912).

Price E. L. and Vandelinde V. D., Robust estimation using the Robbins-Monro stochastic approximation algorithm, IEEE Trans. Information Theory, IT-25 (1979) 698-704.

Prokhorov Y. V., Convergence of random processes and limit theorems in probability theory, Theory of Probab. Appl. 1 (1956) 157-214.

Puri M. L. (Ed.), Nonparametric Techniques in Statistical Inferences, Cambridge Univ. Press, (1970).

Quenouille M. H., Notes on bias in estimation, Biometrika, 43 (1956) 353-360.

Ralston M. L. and Jennrich R. I., DUD, a derivative-free algorithm for nonlinear least-squares, Technometrics, 20 (1978) 7-14.

Ramsay J. O., A comparative study of several robust estimates of slope, intercept and scale in linear regression, J. Amer. Statist. Ass. **72** (1977) 608-615.

Ramsey J. B., Nonlinear estimation and asymptotic approximations, Econometrica, **46** (1978) 901-929.

Reeds J. A., On the definition of von Mises functionals, Ph. D. Dissertation, Harvard Univ., Boston (1976).

Reeds J. A., Jackknifing maximum likelihood estimates, Ann. Statist. **6** (1978) 427-439.

Reid N., Influence functions for censored data, Ann. Statist. **9** (1981a) 78-92.

Reid N., Estimating the median survival time, Biometrika, **68** (1981) 601-608.

Relles D. A. and Rogers W. H., Statisticians are fairly robust estimators of location, J. Amer. Statist. Ass. **72** (1977) 107-111.

Revesz P. (Ed.), Limit Theorems of Probability Theory, North-Holland, (1975).

Rey W., Robust estimates of quantiles, location and scale in time series, Philips Res. Rept, **29** (1974) 67-92.

Rey W., On least pth power methods in multiple regressions and location estimations, BIT, **15** (1975a) 174-184.

Rey W., Mean life estimation from censored samples, Biométrie-Praximétrie **15** (1975b) 145-159.

Rey W., M estimators in robust regression : A case study, in Barra et al. (1977) 591-594.

Rey W. J. J., Robust Statistical Methods, Lecture Notes in Mathematics, Springer-Verlag, **690** (1978).

Rey W. J. J., Robust estimation in descriptive statistics, MBLE Res. Lab., Brussels, **R-385** (1979).

Richardson G. D., Applications of convergence spaces, Bull. Austral. Math. Soc. **21** (1980) 107-123.

Rieder H., Estimates derived from robust tests, Ann. Statist. **8** (1980) 106-115.

Rizzi A. (Edt.), Monte Carlo Studies on Robustness, CISU, Rome.

Robertucci M. L., Bibliography, in Rizzi, (1980) 143-163.

Robinson G. K., Conditional properties of statistical procedures for location and scale parameters, Ann. Statist. **7** (1979) 756-771.

Rocke D. M. and Downs G. W., Estimating the variances of robust estimators of location: influence curve, jackknife, and bootstrap, Commun. Statist. - Simula. Computa. **B-10** (1981) 221-248.

Rocke D. M., Downs G. W. and Rocke A. J., Are robust estimators really necessary? Technometrics, **24** (1982) 95-101.

Ronchetti E., Robust Testing in linear Models: the infinitesimal Approach, Doctoral Thesis, ETH, Zurich (1982).

Ronner A. E., P-norm estimators in a linear regression model, Ph. D. Dissertation, Groningen Univ., Groningen (1977).

Ross G. J., The efficient use of function minimization in non-linear maximum likelihood estimation, Appl. Statist. **19** (1970) 205-221.

Ross G. J. S., Exact and approximate confidence regions for functions of parameters in non-linear models, COMPSTAT 1978, (1978) 110-116.

Ross G. J. S., Uses of non-linear transformation in non-linear optimization problems, COMPSTAT 1980, (1980) 382-388.

Rousseeuw P. J. and Ronchetti E., The influence curve for tests, E. T. H. Zurich, Res. Rept. **21** (1979).

Rousseeuw P. J. and Ronchetti E., Influence curves of general statistics, J. Comp. Appl. Math. **5** (1981) 161-166.

Rubin D. B., The bayesian bootstrap, Ann. Statist. **9**. (1981) 130-134.

Runnenburg J. T., Mean, median, mode, Statistica Neerl. **32** (1978) 1-5.

Ruppert D. and Carroll R. J., Robust regression by trimmed least-squares estimation, Univ. North Carolina, Chapel Hill, **1186** (1978).

Ruppert D. and Carroll R. J., Trimming the least squares estimator in the linear model by using a preliminary estimator, Univ. North Carolina, Chapel Hill, **1220** (1979a).

Ruppert D. and Carroll R. J., M estimates for the heteroscedastic linear model, Univ. North Carolina, Chapel Hill, **1243** (1979b).

Ruppert D. and Carroll R.J., On Bickel's tests for heteroscedasticity, Univ. North Carolina, Chapel Hill, **1244** (1979c).

Ruppert D. and Carroll R. J., Trimmed least squares estimation in the linear model, Univ. North Carolina, Chapel Hill, **1270** (1980).

Safanov M. G. and Athans M., Robustness and computational aspects of nonlinear stochastic estimators and regulators, IEEE Trans. Autom. Contr. **AC-23** (1978) 717-725.

Scarf H., The Computation of Economic Equilibria, Yale Univ. Press, (1973).

Schaefer M., Note on the k-dimensional Jensen inequality, Ann. Probab. **4** (1976) 502-504.

Schoenhage A., Paterson M. and Pippenger N., Finding the median, J. Computer Syst. Sc. **13** (1976) 184-199.

Schrader R. M. and Hettmansperger T. P., Robust analysis of variance based upon the likelihood ratio criterion, Biometrika, **67** (1980) 93-101.

Schucany W. R., Gray H. L. and Owen D. B., On bias reduction in estimation, J. Amer. Statist. Ass. **66** (1971) 524-553.

Sen P. K., Some invariance principles relating to jackknifing and their role in sequential analysis, Ann, Statist. **5** (1977) 316-329.

Sharot T., The generalized jackknife : finite samples and subsample sizes, J. Amer. Statist. Ass. **71** (1976a) 451-454.

Sharot T., Sharpening the jackknife, Biometrika, **63** (1976b) 315-321.

Shorack R., Robust studentization of location estimates, Statistica Neerlandica, **30** (1976) 119-141.

Singh K., On the asymptotic accuracy of Efron's bootstrap, Ann. Statist. **9** (1981) 1187-1195.

Smith V. K., A simulation analysis of the power of several tests for detecting heavy- tailed distributions, J. Amer. Statist. Ass. **70** (1976) 662-665.

Stavig G. R. and Gibbons I. D., Comparing the mean and the median as measures of centrality, Int. Statist. Rev. **45** (1977) 63-70.

Stigler S. M., Simon Newcomb, Percy Daniell and the history of robust estimation: 1895-1920, J. Amer. Statist. Ass. **68** (1973) 872-879.

Stigler S. M., Do robust estimators work with real data? Ann. Statist. **5** (1977) 1055-1098.

/ Swaminathan S. (Ed.), Fixed point theory and its applications, Academic Press (1976).

Sweeting T. J., Speeds of convergence and asymptotic expansions in the central limit theorem: A treatment by operators, Ann. Prob. **8** (1980) 281-297.

Takeuchi K., A survey of robust estimation of location: Models and procedures, especially in case of measurement of a physical quantity, in I.

S. I., **1** (1975) 336-348.

Thorburn D., Some asymptotic properties of jackknife statistics, Biometrika, **63** (1976) 305-313.

Thucydides, History of the Peloponnesian War, **3** (428 B. C.) §20, in Eisenhart (1971).

Tollet I. H., Robust forecasting for the linear model with emphasis on robustness toward occasional outliers, IEEE Int. Conf. Cybernetics and Society, Washington (1976) 600-605.

Todd M. J., The Computation of Fixed Points and Applications, Lecture Notes Econ. Math. Syst., Springer Verlag, **124** (1976).

Tukey J. W., Bias and confidence in not-quite large samples (abstract), Ann. Math. Statist. **29** (1958) 614.

Tukey J. W., A survey of sampling from contaminated distributions, in Olkin (1960) 448-485.

Tukey J. W., Exploratory Data Analysis, Addison-Wesley Publ. Cy., (1977).

Velleman P. F. and Ypelaar M. A., Constructing regressions with controlled features: A method for probing regression performance, J. Amer. Statist. Ass. **75** (1980) 839-844.

von Mises R., On the asymptotic distribution of differentiable statistical functions. Ann. Math. Statist. **18** (1947) 309-348.

von Weber S., A shortened method for the calculation of ranks, Biom. J. **19** (1977) 275-281.

Wahrendorf J., The application of robust non-linear regression methods for fitting hyperbolic Scatchard plots, Int. J. Bio-Medical Computing, **10** (1979) 75-87.

Wahrendorf J. and Brown C. C., A basic inequality in the analysis of the joint action of two drugs, Biometrics, **36** (1980) 653-657.

Walker A. M., On the classical Bonferroni inequalities and the corresponding Galambos inequalities, J. Appl. Prob. **18** (1981) 757-763.

Walley P. and Fine T. L., Towards a frequentist theory of upper and lower probability, Ann. Statist. **10** (1982) 741-761.

Wampler R. H., Solutions to weighted least squares problems by modified Gram-Schmidt with iterative refinements, ACM Trans. Math. Softw. **5** (1979) 457-465.

Weber H. and Werner W., On the accurate determination of nonisolated solutions of nonlinear equations, Computing, **26** (1981) 315-326.

Welsch R. E., Robust and bounded-influence regression, in I. S. I. **2** (1979) 59-68.

Wold H. (Ed.), The Fix-Point Approach to Interdependent Systems, North-Holland, (1981).

Wolfe J. M., On the convergence of an algorithm for discrete Lp approximation, Num. Math. **32** (1979) 439-459.

Woodruff R. S. and Causey B. D., Computerized method for approximating the variance of a complicated estimate, J. Amer. Statist. Ass. **71** (1976) 315-321.

Wretman J., A simple derivation of the asymptotic distribution of a sample quantile, Scand. J. Statist. **5** (1978) 123-124.

Yale C. and Forsythe A. B., Winsorized regression, Technometrics, **18** (1976) 291-300.

Yohai V. J., Robust estimation in the linear model, Ann. Statist. **2** (1974) 562-567.

Yohai V. J. and Maronna R. A., Asymptotic behavior of general M estimates for regression and scale with random carriers, Ann. Statist. **7**
(1979) 258-268.

Youden W. J., Enduring values, Technometrics, **14** (1972) 1-11.

Chapter 13
Subject Index

Universitext

Editors: F. W. Gehring, P. R. Halmos, C. C. Moore

S.-s. Chern

Complex Manifolds Without Potential Theory

(with an appendix on the geometry of characteristic classes)
2nd edition. 1979. V, 152 pages
ISBN 3-540-90422-0
(Originally published by Van Nostrand in 1968)

A. J. Chorin, J. E. Marsden

A Mathematical Introduction to Fluid Mechanics

1979. 85 figures. VII, 205 pages
ISBN 3-540-90406-9

H. Cohn

A Classical Invitation to Algebraic Numbers and Class Fields

with two appendices by O. Taussky
"Artins 1932 Göttingen Lectures on Class Field Theory" and "Connections between Algebraic Number Theory and Integral Matrices"
1978. 19 figures, 3 tables. XIII, 328 pages
ISBN 3-540-90345-3

M. L. Curtis

Matrix Groups

1979. 1 figure. XII, 191 pages
ISBN 3-540-90462-X

D. van Dalen

Logic and Structure

2nd edition. 1983. X, 207 pages
ISBN 3-540-12831-X

K. Devlin

Fundamentals of Contemporary Set Theory

1979. VIII, 182 pages
ISBN 3-540-90441-7

R. E. Edwards

A Formal Background to Mathematics IIa and IIb

A Critical Approach to Elementary Analysis
1980. Part a: XLVII, pages 1–606,
Part b: VI, pages 607–1170
(In 2 parts, not available separately).
ISBN 3-540-90513-8

R. E. Edwards

A Formal Background to Mathematics Ia and Ib

Logic, Sets and Numbers
1979. Part a: XXXIV, pages 1–467,
Part b: IX, pages 468–933
(In 2 parts, not available separately).
ISBN 3-540-90431-X

J. C. Frauenthal

Mathematical Modeling in Epidemiology

1980. IX, 118 pages
ISBN 3-540-10328-7

C. F. Gardiner

A First Course in Group Theory

1980. IX, 227 pages
ISBN 3-540-90545-6

Springer-Verlag
Berlin
Heidelberg
New York
Tokyo

W. Greub

Multilinear Algebra
2nd edition. 1978. VII, 294 pages
ISBN 3-540-90284-8

P. Hájek, T. Havránek

Mechanizing Hypothesis Formation
Mathematical Foundations for a General Theory
1978. XV, 396 pages
ISBN 3-540-08738-9

H. Hermes

Introduction to Mathematical Logic
Translator: D. Schmidt
1973. XI, 242 pages
ISBN 3-540-05819-2

J. G. Kalbfleisch

Probability and Statistical Inference II
1979. 30 figures. IV, 316 pages
ISBN 3-540-90458-1

P. Kelly, G. Matthews

The Non-Euclidean, Hyperbolic Plane
Its Structure and Consistency
1981. 201 figures. XIII, 333 pages
ISBN 3-540-90552-9

A. I. Kostrikin

Introduction to Algebra
Translated from the Russian by N. Koblitz
1982. XIII, 575 pages
ISBN 3-540-90711-4

Y.-C. Lu

Singularity Theory and an Introduction to Catastrophe Theory
3rd corrected printing. 1980. 83 figures. 3 tables.
XII, 199 pages
ISBN 3-540-90221-X

D. A. Marcus

Number Fields
1977. VIII, 279 pages
ISBN 3-540-90279-1

R. M. Meyer

Essential Mathematics for Applied Fields
1979. 34 figures, 2 tables. XVI, 555 pages
ISBN 3-540-90450-6

E. E. Moise

Introductory Problem Courses in Analysis and Topology
1982. VII, 94 pages
ISBN 3-540-90701-7

J. T. Oden, J. N. Reddy

Variational Methods in Theoretical Mechanics
2nd edition. 1983. XI, 309 pages
ISBN 3-540-11917-5

R. B. Reisel

Elementary Theory of Metric Spaces
A Course in Constructing Mathematical Proofs
1982. XI, 120 pages
ISBN 3-540-90706-8

C. E. Rickart

Natural Function Algebras
1979. 2 figures. XIII, 240 pages
ISBN 3-540-90449-2

M. Schreiber

Differential Forms
A Heuristic Introduction
1977. 21 figures. X, 147 pages
ISBN 3-540-90287-2

Springer-Verlag Berlin Heidelberg New York Tokyo

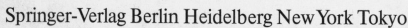